A Jeffrey
University of Newcastle upon Tyne

Quasilinear hyperbolic systems and waves

Pitman Publishing
LONDON · SAN FRANCISCO · MELBOURNE

First published 1976
Reprinted 1977

AMS Subject Classifications: (main) 35–02, 35L60, 35L65
(subsidiary) 73D15, 76J10, 76L05, 76W05

PITMAN PUBLISHING LTD
Pitman House, 39 Parker Street, London WC2B 5PB, UK

PITMAN PUBLISHING CORPORATION
6 Davis Drive, Belmont, California 94002, USA

PITMAN PUBLISHING PTY LTD
Pitman House, 158 Bouverie Street, Carlton, Victoria 3053, Australia

PITMAN PUBLISHING
COPP CLARK PUBLISHING
517 Wellington Street West, Toronto M5V 1G1, Canada

SIR ISAAC PITMAN AND SONS LTD
Banda Street, PO Box 46038, Nairobi, Kenya

PITMAN PUBLISHING CO SA (PTY) LTD
Craighall Mews, Jan Smuts Avenue, Craighall Park,
Johannesburg 2001, South Africa

ISBN 0 273 00102 7

Reproduced and printed by photolithography and bound in Great Britain by Biddles Ltd, Guildford, Surrey

7602:97

To the memory of

my Mother

Preface

There have, in recent years, been many contributions to the study of quasilinear first order hyperbolic systems of equations. These range from the abstract to the particular, and concern both purely mathematical issues and applications to the physical world. As the solution to such systems may be interpreted in terms of waves, which belong to a certain function class and propagate in some suitable space, the work all has as a common feature the fact that it adds to the understanding of what may be called nonlinear wave propagation. To be more precise, when restricting consideration to the hyperbolic case and employing the language of mathematical physics, such nonlinear wave propagation will be both non-dissipative and non-dispersive.

The subject of nonlinear wave propagation is large, and its literature vast, and a newcomer is likely to find difficulty in gathering together in compact form those ideas that are really basic to it. The notes that follow were written with this difficulty in mind and aim to help resolve it within the class of first order quasilinear systems in one space dimension and time, though they also offer some information about systems in many dimensions. A series of naturally related topics is covered with the writing being in part at an expository level and in part at a level appropriate to research. The unifying theme connecting these topics is the effect nonlinearity has in leading to the breakdown of classical differentiable solutions and the way in which, for conservation systems, these solutions may be extended by the introduction of discontinuous solutions which are called shocks. Closely associated with this work is the analysis of the transport of Lipschitz discontinuities in the initial data that characterise wavefronts, and the manner in which they lead to shock formation and interact with established shocks.

The notes themselves derive from a regular course of postgraduate lectures, and from seminars, delivered at Newcastle and elsewhere during the last few years. They are now offered in this more permanent form in the hope that they will be of use to those who wish to enter any of the diverse fields that make demands on quasilinear hyperbolic systems and, by implication, on the ideas set out here. As the notes have been written with physical applications in mind it seemed best to avoid abstraction and to use only classical analysis. The author believes that this approach is appropriate when introducing this material and that it is not unduly restrictive for most applications that are made.

One of the privileges when delivering lectures, or when writing a research note as opposed to a formal monograph, is the right to be selective when choosing material. It is hoped by exercising this right when writing this research note, that the gaps so created are not unduly serious. The main criterion for the selection of material has been that it is useful for the study of nonlinear wave and wavefront propagation in one space dimension and time.

The bibliography makes no pretence at being complete, as it contains only a small fraction of the available literature which is now growing faster than ever before. It should be interpreted merely as a selection of references which the author considers to be helpful and which reflect his particular interest in the subject.

In conclusion, the author wishes to acknowledge with thanks the patient and efficient typing by his secretary Mrs. L. Redpath who has converted rough notes into a finished manuscript.

Newcastle upon Tyne

January 1976

Alan Jeffrey

Contents

1 Nonlinear equations and quasilinear systems

1.1 Waves and Wave Propagation

Those aspects of mathematics which relate to wave propagation occupy a position of considerable importance in the discipline, both from the point of view of the theory of the partial differential equations which characterise wave propagation and because of the numerous and diverse situations under which these equations find application. Probably the most familiar physical notion of a wave, and it is the one that will largely concern us in these notes, is the one involving the propagation of a disturbance that may or may not be of a localised type. This form of wave is thus inherently connected with motion of some kind involving the space $\mathbb{R}^n$ and the time t, so that it gives rise naturally to problems of an evolutionary nature with respect to time. For this reason the time variable will always need to be distinguished from the other independent variables.

When all the equations involved are linear the principle of superposition of solutions applies, and then the extensively developed theory of linear operators can be employed to arrive at solutions in any number of space dimensions and time. An abstract account of this approach is to be found in the works by Hörmander [1] and by Mizohata [2], whilst physically motivated accounts are to be found in the works by Friedman [3], Mikhlin [4], Stakgold [5] and Courant and Hilbert [6]. In the event that the partial differential equations are nonlinear, as in fluid dynamics, or that the field equations are linear but the constitutive equations relating some of the dependent variables are nonlinear, as in nonlinear electrodynamics involving ferromagnetic materials where $\underline{B} = \underline{B}(\underline{H})$, the linear theory is no longer applicable. Under these circumstances different and less general methods of approach must be adopted, when to obtain analytical results it is also often necessary to restrict the number of space dimensions involved. Most frequently this restriction is to systems involving one space dimension and time, for which a fairly complete theory exists, though many useful results of a special nature are also available for more space dimensions. It is mainly this pattern of development of the subject to date that has determined the topics that have been selected for discussion here.

Since the effects of nonlinearity on the time evolution of solutions will be our prime concern, no particular attention will be paid to the other familiar notion of a wave that exists and which is characterised by periodicity of wave pattern. This is because such waves need not necessarily involve movement, as in the case of standing waves. In addition to this, an explicit relationship only exists between such periodic waves and propagating waves when the equations involved are linear, when it takes the form of the Fourier integral theorem.

Although wave propagation can be described by partial differential equations that may be classified either as hyperbolic or as parabolic, it is hyperbolic equations that characterise such propagation most naturally. Indeed, in most physical situations hyperbolic equations provide the basic mathematical approximation to wave propagation, and it is only with the introduction of dissipative effects like viscosity and dispersive effects like those that occur in the Boussinesq equations for water waves [7] that the hyperbolic model becomes parabolic. Because of this importance of hyperbolic equations, and the fact that parabolic equations require a somewhat different method of approach, we shall confine our attention to them and, in particular, to quasilinear first order hyperbolic systems.

The mathematical theory of hyperbolic equations is dominated by the concept of characteristic hypersurfaces and their geometry. The importance of these families of hypersurfaces is not only on account of the fact that they provide a natural coordinate system in which to re-express hyperbolic equations, but also because of the relationship they bear to certain analytical attributes possessed by the solution. To be precise, they are the hypersurfaces across which a continuous solution may exhibit Lipschitz (bounded) discontinuities in its first or higher order normal derivatives. It is essentially on account of this fact that the characteristic hypersurfaces act as transporters of these discontinuities, when they exist, just as they also transport elements of a solution hypersurface when it is differentiable (smooth).

This property of hyperbolic equations is of special significance when related to applications, as may be appreciated most easily when the equations involve only one space dimension and time. In this case the characteristic hypersurfaces reduce to families of characteristic curves in the (x,t)-plane, along each of which may be transported a Lipschitz discontinuity in a derivative of the solution normal to the characteristic. The solution hypersurface itself then reduces to an ordinary smooth surface on which a Lipschitz discontinuity in the first derivative of the solution normal to a characteristic curve manifests itself in the form of a crease on the surface.

When physical problems are associated with such situations this crease in the solution surface, or its analogue in $\mathbb{R}^n \times t$, may be interpreted as representing a clearly defined propagating wavefront. The solution on the side of the wavefront towards which propagation takes place may then be regarded as being the "undisturbed solution" ahead of the wavefront, whilst the solution on the other side may be regarded as a propagating "disturbance wave" which is entering a region occupied by the undisturbed solution. It is this geometrical concept in the context of a physical application that leads to the solution ahead of such a Lipschitz discontinuity sometimes being referred to as the "undisturbed state". This is because the solution at a point in the undisturbed region characterises the state of the physical system at that time and place before the advancing wave has reached it.

When equations in one space dimension and time are involved it is appropriate to refer to the characteristic along which a wavefront propagates as the wavefront trace. This name may be justified by the fact that the characteristic in question is merely the projection of the wavefront onto the (x,t)-plane. Naturally a strictly analogous situation also exists for waves and wavefronts associated with hyperbolic equations defined in $\mathbb{R}^n \times t$.

It will prove convenient to refine the notion of the wavefront a little and also to introduce a notation which identifies the type of Lipschitz discontinuity that is to be studied on the wavefront. This is because it will be necessary to examine the propagation not only of a Lipschitz discontinuity in a first order derivative of the solution normal to the wavefront trace, but also the manner of propagation of higher order Lipschitz discontinuities. For this reason a Lipschitz discontinuity in an n-th order derivative of the solution across a wavefront will be called a C^n discontinuity, not to be confused with the notation $C^{(n)}$ which will be used to denote the n-th family of characteristic curves in the (x,t)-plane.

The refinement of the concept of a wavefront just mentioned comes about when the first Lipschitz discontinuity to occur is one for which $n > 1$. The solution surface (or hypersurface) will then appear smooth across the wavefront, which will coincide with the line in the solution surface across which a discontinuous change of curvature takes place. Thus a wavefront separating an undisturbed state from a propagating disturbance wave will still exist, but it will then be appropriate to call such a disturbance a smooth fronted wave.

In some literature, a solution exhibiting a C^1 discontinuity across a wavefront is said to have a weak discontinuity defining the wavefront. This simpler terminology is useful at times and serves to emphasise the difference between a situation in which the solution itself is continuous across the wavefront (weak and C^n discontinuities), and the case of a strong discontinuity in which the solution is discontinuous across some surface (or hypersurface) whose projection is not a characteristic.

The geometrical interpretation of these notions in the case of a propagating C^1 discontinuity in the solution u to an equation defined in the (x,t)-plane is illustrated in Fig.1.

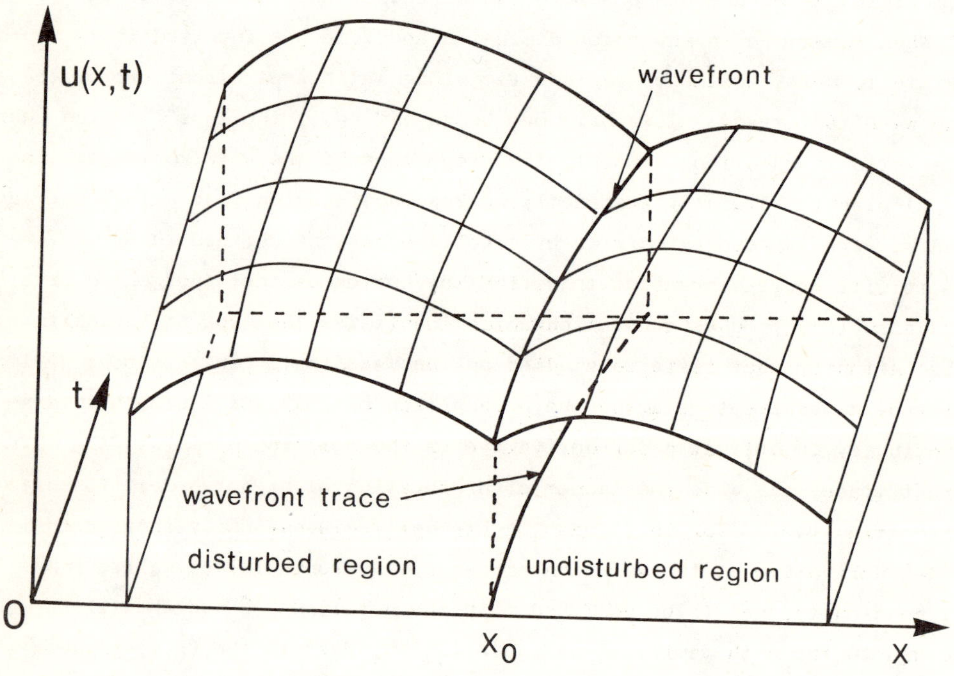

Fig.1. The solution surface, wavefront and wavefront trace with an initial C^1 discontinuity at $u(x_o,0)$.

1.2 First Order Quasilinear Systems and Higher Order Equations

A general partial differential equation in $\mathbb{R}^n$ involving the scalar u and the independent variable vector $\underline{x} = (x_1, x_2, \ldots, x_n)$ is any functional relationship connecting $\underline{x}$, u and the partial derivatives

$$\left(\frac{\partial^{\alpha_1 + \alpha_2 + \ldots + \alpha_n}}{\partial x_1^{\alpha_1} \partial x_2^{\alpha_2} \ldots \partial x_n^{\alpha_n}}\right) u, \tag{1.1}$$

where $(\alpha_1, \alpha_2, \ldots, \alpha_n)$ are n-tuples of non-negative integers. The integer $\alpha_1 + \alpha_2 + \ldots + \alpha_n$ associated with the general partial derivative represented by (1.1) is the order of that derivative, and it is conventional to denote this sum by $|\alpha|$. When this is done, the n-tuples $(\alpha_1, \alpha_2, \ldots, \alpha_n)$ are then called multi-indices, and the appending to (1.1) of the condition that $|\alpha| \leq m$, with m a positive integer, should be interpreted to mean all partial derivatives of u up to and including those of order m.

It is customary to abbreviate the partial derivatives (1.1) by employing the multi-index notation and writing symbolically

$$D^{\alpha} u \equiv \frac{\partial^{|\alpha|} u}{\partial x_1^{\alpha_1} \partial x_2^{\alpha_2} \ldots \partial x_n^{\alpha_n}} . \tag{1.2}$$

This concise notation then allows a general partial differential equation of order m to be written in the compact form

$$F(\underline{x}, u, D^{\alpha} u) = 0; \quad |\alpha| \leq m, \tag{1.3}$$

where F is an arbitrary function of its arguments.

This definition extends immediately to a system of q partial differential equations, of respective orders $m_1, m_2, \ldots, m_q$, involving the p scalar dependent variables $u_1, u_2, \ldots, u_p$. Defining the dependent variable vector $\underline{u} = (u_1, u_2, \ldots, u_p)$ then allows the system to be written

$$\begin{aligned} F_1(\underline{x}, \underline{u}, D^{\alpha} \underline{u}) &= 0 \ ; \ |\alpha| \leq m_1, \\ F_2(\underline{x}, \underline{u}, D^{\beta} \underline{u}) &= 0 \ ; \ |\beta| \leq m_2, \\ &\ldots\ldots\ldots \\ F_q(\underline{x}, \underline{u}, D^{\delta} \underline{u}) &= 0 \ ; \ |\delta| \leq m_q, \end{aligned} \tag{1.4}$$

where apart from the assumption that the q functions F_i are all independent, they are to be regarded as arbitrary functions of their arguments. The order of system (1.4) is max $\{m_1, m_2, \ldots, m_q\}$. The system (1.4) will be said to be a determined system if $p = q$, whilst it will be said to be underdetermined if $q < p$ and overdetermined if $q > p$.

Throughout most of these notes we will work with determined systems, apart from when we consider the reflection and transmission of C^n discontinuities by a strong discontinuity or shock. In these cases the concept of determinacy arises naturally when we examine under what conditions a solution exists and is unique. In general terms, an undetermined system leads to non-uniqueness of solutions whilst an overdetermined system will have no solution unless very special conditions are satisfied.

Partial differential equations of the form (1.3) and systems of the form (1.4) are too ill-defined to permit a detailed study, so that it is necessary to assume some structure for the arbitrary functions that are involved. The simplest structure occurs when the F and F_i are linear in the dependent variables and their derivatives, though this case will not be considered here since our concern will be with the effect of nonlinearity. The next structure, in order of increasing complexity, occurs when the arbitrary functions involved are linear in the highest order derivatives. Systems with this property are said to be quasilinear, and of prime concern to us will be first order quasilinear systems.

The importance of such systems stems from the fact that the mathematical description of many physical situations gives rise to them in a natural manner. Furthermore, many of the higher order scalar equations of mathematical physics are in fact obtained from first order systems after the eliminiation of dependent variables since the structure of the systems is often simple enough to permit this. A familiar example of this type is to be found in Maxwell's equations which form a linear first order system. When either $\underline{E}$ or $\underline{H}$ is eliminated from the system the second order scalar linear wave equation results for each of the components of the remaining vector variable. Similar situations can occur in which quasilinear equations are involved.

This suggests that the connection between a general scalar partial differential equation and a first order quasilinear system should be explored further. Although the elementary result we now demonstrate is well known [8, page 25], its importance in establishing the connection between a general higher order equation and a quasilinear system justifies its repetition.

For simplicity of exposition let us work with a general second order scalar equation in $\mathbb{R}^2$ with independent variable vector $\underline{x} = \underline{x}(x,y)$ and, to conform to the standard notation, let us write $p = u_x$, $q = u_y$, $r = u_{xx}$, $s = u_{xy}$ and $t = u_{yy}$. The general partial differential equation then assumes the form

$$F(x, y, u, p, q, r, s, t) = 0, \tag{1.5}$$

and it may, or may not, be quasilinear. Suppose first that it is not quasilinear, but that F is suitably differentiable, so that we may compute the total derivative of F with respect to y. It then follows that after differentiation (1.5) yields the result

$$F_r r_y + F_s s_y + F_t t_y + F_p p_y + F_q q_y + F_u q + F_y = 0. \tag{1.6}$$

This equation is a first order quasilinear equation in terms of the variables u, p, q, r, s and t, though as it stands (1.6) is underdetermined. The situation may be easily remedied by associating with (1.6) the following five first order equations

$$q = u_y, \quad s = p_y, \quad t = q_y, \quad s_x = r_y \quad \text{and} \quad s_y = t_x, \tag{1.7}$$

when the combined system (1.6) and (1.7) becomes a properly determined system of six equations from which to obtain u, p, q, r, s and t. The first three equations in (1.7) follow directly from the definitions of p, q, s and t, whilst the last two follow from the equality of mixed derivatives which we effectively pre-suppose when deducing (1.6).

A solution of system (1.6) and (1.7) (and indeed of (1.5)) will be taken to be a function u that satisfies the equation (or the system) and also satisfies some auxiliary conditions comprising the initial or boundary conditions. Naturally the determination of suitably posed conditions is connected with the type of equations (hyperbolic or elliptic etc.) that are involved, but for our present purpose let it suffice to suppose that it is proper that u and u_y be assigned arbitrarily on the "initial" line $y = y_0$.

It is then a simple matter to use this information together with (1.7) to determine all the "initial" values for the system along the line $y = y_0$, with the exception of t which can then be found from (1.5) itself provided the implicit relation (1.5) can be solved for t. To prove that a solution of the system (1.6) and (1.7) is also a solution of (1.5), it is first necessary to show that when the initial conditions for (1.5) are applied to the system they lead to identities of the form $r \equiv u_{xx}$ etc., on $y = y_o$. The proof that a solution of the system is also a solution of (1.5) will then be complete once it is established that $dF/dy = 0$; for as initially $F = 0$, it will then follow that u satisfies (1.5).

As this last step in the proof is immediate, we only indicate how the first results may be established. We have at once that

$$r_y = s_x = p_{yx} = p_{xy},$$

so that

$$(r - p_x)_y = 0.$$

However, $r = p_x$ for $y = y_o$, so that we have shown $r \equiv p_x \equiv u_{xx}$ along $y = y_o$. Similar arguments lead to the other identities that are required. In the event that (1.5) is quasilinear, and not a general equation as supposed, the initial computation of dF/dy could have been omitted with a consequent simplification of the argument. The fact solutions of (1.5) subject to the initial values on $y = y_o$ are also solutions of the system with its corresponding initial values is obvious.

Although a general equation of the form (1.5) may always be reduced to the system of six equations (1.6) and (1.7), such a reduction is not unique, and in special cases it may be accomplished with fewer equations. A case in point is the quasilinear second order equation,

$$u_{yy} = c^2(1 + ku_x)u_{xx}, \tag{1.8}$$

in which u is not present explicitly. In terms of $p = u_x$, $q = u_y$ this may be expressed as the first order quasilinear system of equations

$$q_y = c^2(1 + kp)p_x = 0 \tag{1.9a}$$

$$p_y - q_x = 0, \tag{1.9b}$$

though once this system has been solved for p and q a further quadrature is required before u can be determined. Alternative reductions of (1.8) to different pairs of equations follow from other changes of variable as, for example, when v, w are defined by the expressions $v = u_x + u_y$ and $w = u_x - u_y$.

Only in such special cases can an n-th order general equation be reduced to a first order quasilinear system comprising n equations though, as just established, a reduction to a larger system is always possible. The converse is not true, in the sense that a first order quasilinear system cannot, in general, be reduced to one equivalent equation of higher order.

Despite the fact that the solution of system (1.6) and (1.7) has been constrained to be the same as that of the general second order equation (1.5), there is a significant difference between the two different formulations of the problem. The second order equation from which we started has two families of real characteristic curves when it is hyperbolic (if $F_s^2 \geq 4F_rF_t$), whilst we shall see later that the first order system has six. In fact the identity of the two solutions has been achieved by causing four of the six families of characteristic curves of the system to become coincident, whilst the remaining two families coincide with the two families associated with (1.5).

1.3 Matrix Formulation of Systems

It is both natural and convenient to work with first order quasilinear systems of equations expressed in matrix form. The most general properly determined system of such equations in $\mathbb{R}^n \times t$ may be written

$$A_o(U,\underline{x},t)U_t + \sum_{i=1}^{n} A_i(U,\underline{x},t)U_{x_i} + B(U,\underline{x},t) = 0, \tag{1.10}$$

where $U(\underline{x},t)$ is a column vector with the m elements $u_1(\underline{x},t)$, $u_2(\underline{x},t)$, $\ldots,u_m(\underline{x},t)$, $\underline{x} = (x_1,x_2,\ldots,x_n)$ is a vector in $\mathbb{R}^n$, the $A_i(U,\underline{x},t)$ are $m \times m$ matrices with elements dependent upon U, $\underline{x}$ and t, $B(U,\underline{x},t)$ is a column vector with the m elements $b_1(U,x,t)$, $b_2(U,\underline{x},t),\ldots,$ $b_m(U,\underline{x},t)$ and the suffixes t and x_i denote partial differentiation with respect to t and x_i, respectively.

On occasions $m \times m$ matrices in the system (1.10) either possesses symmetry or they can be made symmetric by pre-multiplication by a suitable matrix. When either all the matrices $A_i(U,\underline{x},t)$ are symmetric, or they may all be made symmetric by pre-multiplication by the same matrix, the system (1.10) will be said to be a symmetric system. A useful sub-class of such systems is the class of symmetric hyperbolic systems that will be defined later. The importance of this class is related both to the properties of symmetric matrices and to the occurrence of such systems in connection with various physical situations.

It is often the case that the system (1.10) may be re-cast into the structurally simpler form

$$F_t + \text{div } G = H, \tag{1.11}$$

where $F = F(U(\underline{x},t))$ and $H = H(U(\underline{x},t))$ are m element column vectors and $G = G(U(\underline{x},t))$ is an $m \times n$ matrix, all with elements dependent upon the elements of U and, through them, on $\underline{x}$ and t. Here, when G is an $m \times n$ matrix, if the s-th column of G is denoted by $g^{(s)}$, then

$$\text{div } G = \sum_{s=1}^{n} \frac{\partial g^{(s)}}{\partial x_x}$$

On account of the fact that system (1.11) is written in divergence form it will, by analogy with physical laws, be said to be expressed in generalised conservation form. It is in strict conservation form when F and G are functions only of U and H = 0, for a non-zero H corresponds to the occurrence of either generalised sources or sinks.

The connection between (1.11) and (1.10) is immediate when the indicated differentiations in (1.11) are performed to give

$$(\nabla_u F)U_t + \sum_{s=1}^{n} (\nabla_u g^{(s)})U_{x_s} = H, \tag{1.12}$$

where $\nabla_u \equiv (\partial/\partial u_1, \partial/\partial u_2, \ldots, \partial/\partial u_m)$ is the gradient operator with respect to the m elements of U. The matrices $(\nabla_u F)$ and $(\nabla_u g^{(s)})$ are thus the Jacobian matrices

$$(\nabla_u F) \equiv [\partial f_i/\partial u_j] \text{ and } (\nabla_u g^{(s)}) \equiv [\partial g_i^{(s)}/\partial u_j],$$

where f_i and $g_i^{(s)}$ are the i-th elements of the column vectors F and $g^{(s)}$, respectively.

Example 1.

A physical problem involving four strict physical conservation laws

One of the simplest examples of a first order quasilinear system is provided by unsteady isentropic compressible gas flow [7,8,9], when the system becomes the following involving four dependent variables

$$\frac{\partial\rho}{\partial t} + u\frac{\partial\rho}{\partial x} + v\frac{\partial\rho}{\partial y} + w\frac{\partial\rho}{\partial z} + \rho\left(\frac{\partial u}{\partial x} + \frac{\partial v}{\partial y} + \frac{\partial w}{\partial z}\right) = 0, \qquad (1.13a)$$

$$\frac{\partial u}{\partial t} + u\frac{\partial u}{\partial x} + v\frac{\partial u}{\partial y} + w\frac{\partial u}{\partial z} + \frac{1}{\rho}\frac{\partial p}{\partial x} = 0, \qquad (1.13b)$$

$$\frac{\partial v}{\partial t} + u\frac{\partial v}{\partial x} + v\frac{\partial v}{\partial y} + w\frac{\partial v}{\partial z} + \frac{1}{\rho}\frac{\partial p}{\partial y} = 0, \qquad (1.13c)$$

$$\frac{\partial w}{\partial t} + u\frac{\partial w}{\partial x} + v\frac{\partial w}{\partial y} + w\frac{\partial w}{\partial z} + \frac{1}{\rho}\frac{\partial p}{\partial z} = 0, \qquad (1.13d)$$

where ρ is the gas density, $\underline{q} = (u, v, w)$ is the gas velocity with components expressed in Cartesian form and $p = p(\rho)$ is the constitutive relation determining the gas pressure. If the sound speed a is introduced via the relation $a^2 = \partial p/\partial\rho$, then the system (1.13) may be expressed in the matrix form

$$U_t + A_1U_x + A_2U_y + A_3U_z = 0, \qquad (1.14a)$$

where

$$U = \begin{bmatrix} \rho \\ u \\ v \\ w \end{bmatrix}, \quad A_1 = \begin{bmatrix} u & \rho & 0 & 0 \\ \frac{a^2}{\rho} & u & 0 & 0 \\ 0 & 0 & u & 0 \\ 0 & 0 & 0 & u \end{bmatrix}, \quad A_2 = \begin{bmatrix} v & 0 & \rho & 0 \\ 0 & v & 0 & 0 \\ \frac{a^2}{\rho} & 0 & v & 0 \\ 0 & 0 & 0 & v \end{bmatrix}$$

$$\text{and } A_3 = \begin{bmatrix} w & 0 & 0 & \rho \\ 0 & w & 0 & 0 \\ 0 & 0 & w & 0 \\ \frac{a^2}{\rho} & 0 & 0 & w \end{bmatrix} \qquad (1.14b)$$

This system may also be expressed in strict conservation form involving physical quantities once it is recognised that equation (1.13a) expresses conservation of mass as it stands, whilst equations (1.13b, c,d) are each related to the scalar equations expressing conservation of momentum in the x, y and z-directions, respectively.

We have immediately that (1.13a) may be written

$$\frac{\partial\rho}{\partial t} + \text{div}(\rho u, \rho v, \rho w) = 0, \qquad (1.15a)$$

whilst combining (1.13a) with (1.13b,c,d) leads to the result

$$\frac{\partial}{\partial t}\begin{bmatrix}\rho u\\ \rho v\\ \rho w\end{bmatrix} + \operatorname{div}\begin{bmatrix}p+\rho u^2 & \rho uv & \rho uw\\ \rho uv & p+\rho v^2 & \rho vw\\ \rho uw & \rho vw & p+\rho w^2\end{bmatrix} = 0. \tag{1.15b}$$

Identifying (1.15a, b) with (1.11), we see that these equations are in strict conservation form with

$$F = \begin{bmatrix}\rho\\ \rho u\\ \rho v\\ \rho w\end{bmatrix}, \quad G = \begin{bmatrix}\rho u & \rho v & \rho w\\ p+\rho u^2 & \rho uv & \rho uw\\ \rho uv & p+\rho v^2 & \rho vw\\ \rho uw & \rho vw & p+\rho w^2\end{bmatrix} \quad \text{and } H = 0.$$

As already remarked, the conservation laws involved here are the ones for mass and momentum.

Example 2.

A physical problem involving two generalised conservation laws

The system of two first order quasilinear equations

$$u_t + uu_x + 2cc_x - H_x = 0, \tag{1.16a}$$

$$2c_t + 2uc_x + cu_x = 0, \tag{1.16b}$$

describes the one-dimensional flow of water in terms of the so-called shallow water approximation, which may be more properly described as the long wave approximation [7, 8, 10]. The two dependent variables involved are u the horizontal component of the water velocity and c the speed of propagation of a surface disturbance, whilst $y + Y(x) = 0$ is the equation of the sea-bed relative to an origin located on the equilibrium surface of the water along which lies the x-axis, whilst $H(x) = gY(x)$ with g the acceleration due to gravity.

In matrix notation the system becomes

$$U_t + AU_x + B = 0, \tag{1.17a}$$

where

$$U = \begin{bmatrix}u\\ c\end{bmatrix}, \quad A = \begin{bmatrix}u & 2c\\ c/2 & u\end{bmatrix} \text{ and } B = \begin{bmatrix}-H_x\\ 0\end{bmatrix}. \tag{1.17b}$$

It is easily seen that equations (1.16) may be expressed in the generalised conservation form

$$\frac{\partial}{\partial t}\begin{bmatrix} u \\ c^2 \end{bmatrix} + \frac{\partial}{\partial x}\begin{bmatrix} \frac{1}{2}u^2 + c^2 - H(x) \\ uc^2 \end{bmatrix} = 0, \qquad (1.18a)$$

from which it follows by comparison with (1.11) that

$$F = \begin{bmatrix} u \\ c^2 \end{bmatrix} \quad \text{and} \quad G = \begin{bmatrix} \frac{1}{2}u^2 + c^2 - H(x) \\ uc^2 \end{bmatrix}. \qquad (1.18b)$$

The mathematically conserved quantities involved here are non-physical, so that although a physical problem is involved, no special physical significance may be attached to the conservation form (1.18). This simple system will be used later to provide various useful examples with which to illustrate the application of general results obtained in these notes.

Example 3

A non-conservative system of three equations without physical significance

The example offered here involves a system of three equations that is of significant mathematical interest, but which has no physical significance [11]. The column vector U_μ with elements $u_{1\mu}$, $u_{2\mu}$, $u_{3\mu}$, depending on a real parameter μ, is required to satisfy the equation

$$\frac{\partial U_\mu}{\partial t} + A(\mu, U_\mu)\frac{\partial U_\mu}{\partial x} = 0, \qquad (1.19a)$$

where

$$A(\mu, U_\mu) = \begin{bmatrix} -\cosh 2\mu u_{2\mu} & 0 & -\sinh 2\mu u_{2\mu} \\ \cosh \mu u_{2\mu} & 0 & \sinh \mu u_{2\mu} \\ \sinh 2\mu u_{2\mu} & 0 & \cosh 2\mu u_{2\mu} \end{bmatrix}. \qquad (1.19b)$$

As this system cannot be expressed in divergence form it does not represent a set of conservation laws. Later we shall see that its solution has some unusual and interesting properties. In particular, we shall see that when certain initial conditions are given a finite solution does not exist for all t.

Example 4

A physical problem involving four strict non-physical conservation laws

The final example has been taken from solid mechanics, and the system of equations involved describes the propagation of one-dimensional waves in non-linear elastic materials [12, 13, 14, 15]. The first order quasilinear system of four equations that is involved is, in fact, obtained from two coupled second order equations of motion for the material by means of the reduction process described in Section 1.2.

If the Cartesian coordinates of a particle of the material in its initial state are denoted by X_i, and the coordinates at time t by x_i, with i = 1,2,3, then the displacement field involved may be written

$$x_i = x_i\,(X_j, t) \quad \text{with } i,\ j = 1,\ 2,\ 3.$$

Considering now the case of plane deformation, we denote by u_α, with $\alpha = 2,3$, the displacement of a particle in the X_α-direction resulting from wave propagation in the X_1-direction, when it follows that

$$x_1 = X_1, \quad x_2 = X_2 + u_2(X_1, t), \quad x_3 = X_3 + u_3(X_1, t).$$

In terms of these coordinates the reduction of the equations of motion involving the displacements u_2, u_3 to a first order system leads to the following result [15]

$$U_t + \mathcal{E}(U)U_{X_1} = 0 \tag{1.20a}$$

with

$$U = \begin{bmatrix} v_2 \\ v_3 \\ p_2 \\ p_3 \end{bmatrix} \quad \text{and} \quad \mathcal{E}(U) = \begin{bmatrix} 0 & 0 & -A_{22} & -A_{23} \\ 0 & 0 & -A_{23} & -A_{33} \\ -1 & 0 & 0 & 0 \\ 0 & -1 & 0 & 0 \end{bmatrix}, \tag{1.20b}$$

where the A_{ij} depend only on the components p_α of U and

$$v_\alpha = \frac{\partial u_\alpha}{\partial t}, \quad p_\alpha = \frac{\partial u_\alpha}{\partial X_1}, \quad \text{for } \alpha = 2,3. \tag{1.20c}$$

The precise form of the elements A_{ij} given in [15] is not relevant at the present stage of the discussion since it is sufficient here that their dependence on the p_α makes system (1.20a) quasilinear. Although the original coupled second order equations had physical significance, in that they were equations of motion, the equivalent strict conservation laws that may be derived from (1.20a) do not, as the system has been obtained by the reduction process described in Section 1.2. As with the form of the elements A_{ij}, so the precise form of the conservation equations expressed in terms of the shear stresses is irrelevant at this stage of the discussion.

1.4. Characteristics and the Cauchy Problem for a General Nonlinear First Order Equation

Although the ideas are classical, and are to be found elsewhere [6,8], it will be useful to begin our examination of wave propagation by presenting a resumé of the concepts underlying the Cauchy problem. We offer this in the context of the general nonlinear first order partial differential equation in $\mathbb{R}^n$ for the function $u(\underline{x})$, with $\underline{x} = (x_1, x_2, \ldots, x_n)$. This setting, although restricted to one dependent variable, is still very general and serves to introduce characteristics, wavefronts and the complicated geometry that is possible for integral surfaces.

From the remarks in Section 1.2 it is apparent that a general nonlinear first order partial differential equation for u will be of the form

$$F(\underline{x}, u, \underline{p}) = 0, \tag{1.21}$$

where $\underline{p} = (p_1, p_2, \ldots, p_n)$, with $p_i = \partial u/\partial x_i$, and F is an arbitrary function of its arguments. An important special case of (1.21) is, of course, when F is linear in the p_i so that the equation becomes quasilinear.

Now let $u = f(\underline{x})$ be a solution of (1.21), and define the scalar function $G(\underline{x}, u)$ of (n+1) variables by the relation

$$G(\underline{x}, u) \equiv f(\underline{x}) - u\,. \tag{1.22}$$

Then $G(\underline{x}, u) =$ const. defines a manifold in (n+1)-dimensions in the space $\mathbb{R}^n \times u$ and the manifold $\mathcal{S}$ corresponding to $G(\underline{x}, u) = 0$ is the solution manifold, or integral surface of (1.21), as it is frequently called. At any point $P(\underline{x}, u)$ of $\mathcal{S}$, the normal $\underline{\nu}$ with elements which are the direction ratios $\nu_1, \nu_2, \ldots, \nu_n$ and ν_u relative to the axes $x_1, x_2, \ldots, x_n$ and u, respectively, is proportional to grad G, so that we may immediately write the (n+1) equations

$$\frac{\nu_1}{\partial G/\partial x_1} = \frac{\nu_2}{\partial G/\partial x_2} = \dots = \frac{\nu_n}{\partial G/\partial x_n} = \frac{\nu_u}{\partial G/\partial u} \ . \tag{1.23}$$

Using the form of $G(\underline{x},u)$ given in (1.22) together with the definition of p_i enables (1.23) to be re-written as

$$\frac{\nu_1}{p_1} = \frac{\nu_2}{p_2} = \dots = \frac{\nu_n}{p_n} = \frac{\nu_u}{-1} = \alpha(\text{say}). \tag{1.24}$$

In view of these relations, we see that (1.21) may be interpreted as being an algebraic constraint on the normal $\underline{\nu}$ to $\mathcal{S}$ at P in order that $G(\underline{x},u) = 0$ should be an integral surface of (1.21). That is to say, (1.24) imposes conditions on the geometry of $\mathcal{S}$ at P.

Expressed differently, (1.21) requires the normal $\underline{\nu}$ to $\mathcal{S}$ at a point P to lie along the generator of the one-parameter family of straight lines defined by (1.24) in terms of the parameter α. The set of equations (1.24) defines a generalised cone with its vertex at P, and thus $\underline{\nu}$ may lie along any one of its generators. In consequence of the one parameter family of normals $\underline{\nu}$ to $\mathcal{S}$ that can exist at P, the geometry of an integral surface in $\mathbb{R}^n \times u$ is obviously more complicated than is likely to occur in $\mathbb{R}^2 \times u$.

At any point P of $\mathcal{S}$ there will be a hyperplane normal to each $\underline{\nu}$ at P which it is convenient to call a tangent hyperplane at P. These hyperplanes are "tangent" in the sense that they are tangent to the integral surface at P. Since the normals $\underline{\nu}$ at P lie along the generators of the cone at P with parameter α, the tangent hyperplanes at P will also envelop a cone at P. This cone is called the Monge cone with vertex at P and, by virtue of its definition, the Monge cone, or one of its generators through P, will be tangent to $\mathcal{S}$ in the vicinity of P. When the generators of the Monge cone have been determined at a point P of the integral surface, and the behaviour of the associated tangent hyperplanes is known along the generators, then the solution of (1.21) is known along the generators. This follows because at any point of a generator the (2n + 1) quantities $x_1, x_2, \dots, x_n$, u, $p_1, p_2, \dots, p_n$ will be known specifying, respectively, a position $\underline{x}$ in $\mathbb{R}^n$, the value of u at that point and the orientation of a normal $\underline{\nu} \propto \underline{p}$, and hence a tangent hyperplane to the solution manifold at that point. We shall have more to say about this idea later. As P was any point of $\mathcal{S}$ a cone of this type, sometimes degenerate, will exist at all points of $\mathcal{S}$.

The generators of the Monge cone will also depend on the parameter α, and from (1.24) we may write $p_i = p_i(\alpha)$, so that (1.21) becomes

$$F(\underline{x}, u, \underline{p}(\alpha)) = 0. \tag{1.25}$$

If σ depends on distance along a generator of the Monge cone measured from a suitable origin on that generator, so that $x_i = x_i(\sigma)$, then along that generator the integral surface $G(\underline{x},u) = 0$ reduces to

$$f(\underline{x}(\sigma)) - u(\sigma) = 0.$$

The solution will then evolve as a function of σ and if, for convenience, σ is considered to be strictly monotonic increasing, the solution can be said to evolve away from the "initial" conditions prevailing at the origin chosen for σ along that generator. Differentiation with respect to σ, coupled with the definition of p_i, then gives the result

$$\sum_{i=1}^{n} p_i(\alpha)\frac{dx_i}{d\sigma} - \frac{du}{d\sigma} = 0 \equiv \Psi(\underline{x},u,\alpha). \tag{1.26}$$

However, the generators of the Monge cone will also be determined as the envelope of the tangent hyperplanes defined by (1.26) in terms of the parameter α. Consequently, employing the elementary theory of envelopes, we see that this is equivalent to solving simultaneously $\Psi = 0$ and $d\Psi/d\alpha = 0$. That is, we solve simultaneously (1.26) and

$$\sum_{i=1}^{n} \frac{dp_i}{d\alpha}\frac{dx_i}{d\sigma} = 0 \ (= \frac{d\Psi}{d\alpha}). \tag{1.27}$$

To proceed further it is now necessary to utilise (1.25), which after differentiation with respect to α yields,

$$\sum_{i=1}^{n} F_{p_i}\frac{dp_i}{d\alpha} = 0. \tag{1.28}$$

Direct comparison of (1.27) and (1.28) then gives

$$\frac{dx_i}{d\sigma} = kF_{p_i} \qquad \text{for } i = 1, 2, \ldots, n$$

with k an arbitrary constant of proportionality. For convenience we set $k = 1$ to obtain

$$\frac{dx_i}{d\sigma} = F_{p_i} \quad \text{for } i = 1, 2, \ldots, n, \tag{1.29}$$

which taken together with (1.26) leads to the result

$$\frac{du}{d\sigma} = \sum_{i=1}^{n} p_i F_{p_i}. \tag{1.30}$$

Equations (1.29) and (1.30) comprise (n+1) equations from which to determine the (2n + 1) quantities $\underline{x}(\sigma)$, $u(\sigma)$ and $\underline{p}(\sigma)$ that we need to know along a generator in order to know the behaviour of the integral surface in the neighbourhood of the generator. A further n equations are required for $dp_i/d\sigma$ in order that $\underline{p}(\sigma)$ may be found. These may be determined from (1.21) once we write $dp_i/d\sigma$ in the form

$$\frac{dp_i}{d\sigma} = \sum_{j=1}^{n} \frac{\partial p_i}{\partial x_j} \frac{dx_j}{d\sigma} . \tag{1.31}$$

Differentiating (1.21) partially with respect to x_i yields

$$F_{x_i} + F_u p_i + \sum_{j=1}^{n} F_{p_j} \frac{\partial p_j}{\partial x_i} = 0 \quad \text{for } i = 1,2,\ldots,n. \tag{1.32}$$

However, as $\partial p_j/\partial x_i = \partial p_i/\partial x_j$, from (1.29) and (1.32) we have

$$F_{x_i} + F_u p_i + \sum_{j=1}^{n} \frac{dx_j}{d\sigma} \frac{\partial p_i}{\partial x_j} = 0,$$

so that employing (1.31) we find

$$\frac{dp_i}{d\sigma} = -(F_{x_i} + F_u p_i) \quad \text{for } i = 1,2,\ldots,n. \tag{1.33}$$

This, then, is the further set of n equations that was required from which $\underline{p}(\sigma)$ may be determined along a generator. Hence there are now (2n + 1) ordinary differential equations (1.29), (1.30) and (1.33) that must be solved if $\underline{x}(\sigma)$, $u(\sigma)$ and $\underline{p}(\sigma)$ are to be found. These are called the characteristic equations of (1.21) and they have as their integral F = const. By employing these geometrical concepts it has thus been possible to reduce the solution of partial differential equation (1.21) to the solution of a simultaneous system of ordinary differential equations. To solve this system initial conditions are required for $\underline{x}$, u and $\underline{p}$, so that it is now necessary to see if this data is compatible with the data it is appropriate to specify for (1.21).

The characteristic equations comprise a first order system of ordinary differential equations determining $u(\sigma)$ and $\underline{p}(\sigma)$ as functions of σ along the trajectories $\underline{x} = \underline{x}(\sigma)$ emanating from points on some initial (n-1)-dimensional sub-manifold of $\mathbb{R}^n$ on which u is specified. To formalise this idea let $\sum^{o}$, this initial sub-manifold, be given in the parametric form

$$x_i = X_i(\beta_1,\beta_2,\ldots,\beta_{n-1}) \text{ for } i = 1,2,\ldots,n, \tag{1.34}$$

and let

$$u = U(\beta_1,\beta_2,\ldots,\beta_{n-1}). \tag{1.35}$$

Then as $x_i = x_i(\sigma)$ and $u = u(\sigma)$, it will be convenient if on each generator we choose the origin for σ to coincide with the point on the initial sub-manifold at which that trajectory or generator $\underline{x} = \underline{x}(\sigma)$ intersects the sub-manifold. Thus the initial conditions for our system of ordinary differential equations, for which the general x_i and u are

$$x_i = x_i(\beta_1,\beta_2,\ldots,\beta_{n-1};\ \sigma) \text{ and } u = u(\beta_1,\beta_2,\ldots,\beta_{n-1};\ \sigma),$$

becomes

$$x_i(\beta_1,\beta_2,\ldots,\beta_{n-1};\ 0) = X_i(\beta_1,\beta_2,\ldots,\beta_{n-1}) \quad \text{for } i = 1,2,\ldots,n \tag{1.36}$$

and

$$u(\beta_1,\beta_2,\ldots,\beta_{n-1};\ 0) = U(\beta_1,\beta_2,\ldots,\beta_{n-1}). \tag{1.37}$$

The determination of the solution of (1.21) which assumes the functional form (1.37) on an initial sub-manifold (1.36) is called the Cauchy problem for the partial differential equation (1.21). The function $U(\beta_1,\beta_2,\ldots,\beta_{n-1})$ is then called the Cauchy data for the problem.

Now the only initial condition specified for the solution u is the Cauchy data (1.37), but when deriving the characteristic equations, the set of equations (1.33) entered naturally determining $p_i = p_i(\beta_1, \beta_2, \ldots, \beta_{n-1}; \sigma)$. Consequently it is also necessary to specify Cauchy data for $\underline{p}$ when $\sigma = 0$. The question that now arises is under what conditions, by using the characteristic equations together with (1.34) and (1.35), it is possible to determine the initial functions.

$$p_i(\beta_1, \beta_2, \ldots, \beta_{n-1}; 0) = P_i(\beta_1, \beta_2, \ldots, \beta_{n-1}) \tag{1.38}$$

for $i = 1, 2, \ldots, n$.

To resolve this final problem we first observe that the Cauchy data $\underline{X}$, U, $\underline{P}$, must satisfy the original partial differential equation

$$F(\underline{X}, U, \underline{P}) = 0, \tag{1.39}$$

whilst from (1.30) it also follows that

$$\left.\frac{dU}{d\sigma}\right|_{\sigma=0} = \sum_{i=1}^{n} P_i \left.\frac{dX_i}{d\sigma}\right|_{\sigma=0}. \tag{1.40}$$

After allowing for the dependence of U and X_i on the β_s this becomes

$$\sum_{s=1}^{n-1} \frac{dU}{d\beta_s} \left.\frac{d\beta_s}{d\sigma}\right|_{\sigma=0} = \sum_{i=1}^{n} \sum_{s=1}^{n-1} P_i \frac{dX_i}{d\beta_s} \left.\frac{d\beta_s}{d\sigma}\right|_{\sigma=0}, \tag{1.41}$$

from which there then follow the (n-1) equations

$$\frac{dU}{d\beta_s} = \sum_{i=1}^{n} P_i \frac{dX_i}{d\beta_s} \quad \text{for } s = 1, 2, \ldots, n-1. \tag{1.42}$$

Thus, between (1.39) and (1.42), there are n equations from which to solve for the n functions P_i that are required to complete the Cauchy data to be specified on the initial sub-manifold defined by (1.34). By virtue of the implicit function theorem the P_i may be determined on the initial sub-manifold provided the Jacobian of the transformation $J = \partial(\underline{X})/\partial(\underline{\beta}, \sigma)$ is non-singular. This Jacobian expressing a condition on equations (1.39) and (1.42) is easily seen to have the following form

$$J = \frac{\partial(X_1, X_2, \ldots, X_n)}{\partial(\beta_1, \beta_2, \ldots, \beta_{n-1}, \sigma)} = \begin{vmatrix} \frac{\partial X_1}{\partial \beta_1} & \frac{\partial X_1}{\partial \beta_2} & \cdots & \frac{\partial X_1}{\partial \beta_{n-1}} & \frac{\partial X_1}{\partial \sigma} \\ \frac{\partial X_2}{\partial \beta_1} & \frac{\partial X_2}{\partial \beta_2} & \cdots & \frac{\partial X_2}{\partial \beta_{n-1}} & \frac{\partial X_2}{\partial \sigma} \\ \cdot & \cdot & \cdot & \cdot & \cdot \\ \frac{\partial X_n}{\partial \beta_1} & \frac{\partial X_n}{\partial \beta_2} & \cdots & \frac{\partial X_n}{\partial \beta_{n-1}} & \frac{\partial X_n}{\partial \sigma} \end{vmatrix} \neq 0. \qquad (1.43)$$

However, on the initial sub-manifold equations (1.29) become $dX_i/d\sigma = F_{P_i}$, so that (1.43) assumes the form

$$J = \begin{vmatrix} \frac{\partial X_1}{\partial \beta_1} & \frac{\partial X_1}{\partial \beta_2} & \cdots & \frac{\partial X_1}{\partial \beta_{n-1}} & F_{P_1} \\ \frac{\partial X_2}{\partial \beta_1} & \frac{\partial X_2}{\partial \beta_2} & \cdots & \frac{\partial X_2}{\partial \beta_{n-1}} & F_{P_2} \\ \cdot & \cdot & \cdot & \cdot & \cdot \\ \frac{\partial X_n}{\partial \beta_1} & \frac{\partial X_n}{\partial \beta_2} & \cdots & \frac{\partial X_n}{\partial \beta_{n-1}} & F_{P_n} \end{vmatrix} \neq 0. \qquad (1.44)$$

The condition $J \neq 0$ corresponds to a restriction on the local geometry at each point of the initial sub-manifold in order that Cauchy data for (1.21) should provide adequate initial data for the characteristic equations. Thus if $J \neq 0$, then for $\sigma = \sigma_1 (> 0)$, equations (1.29), (1.30) and (1.33) serve to define a new Cauchy problem on the sub-manifold $\underline{x} = \underline{x}(\sigma_1)$. If $J \neq 0$ at points of this new sub-manifold then the solution can be extended further. Points Q_o on the initial sub-manifold $\sum^o$ in $\mathbb{R}^n$ corresponding to $\sigma = 0$ will map to points Q_σ on the sub-manifold $\sum^\sigma$ corresponding to $\sigma > 0$ along the trajectories $\underline{x} = \underline{x}(\sigma)$. These trajectories are called the rays or bi-characteristics of the partial differential equation (1.21). In general we may conclude that, aside from the occurrence of geometrical singularities in the sub-manifold $\sum^\sigma$, if the rays through each point Q_o of the initial sub-manifold $\sum^o$ can be constructed, then so can the solution. In physical problems these singularities, when they occur, coincide with the occurrence of interesting physical phenomena.

Let us look briefly at the geometrical and analytical implications of the condition $J \neq 0$, and also at the associated condition $J = 0$. As the general Monge cone is defined by (1.29) in the form

$$\frac{dx_1}{F_{p_1}} = \frac{dx_2}{F_{p_2}} = \cdots = \frac{dx_n}{F_{p_n}} = d\sigma, \tag{1.45}$$

let us examine its form on the initial sub-manifold $\sum^o$ defined by (1.34). If displacements constrained to $\sum^o$ are denoted by dx_i^o, then (1.45) becomes

$$\frac{dx_1^o}{F_{p_1}} = \frac{dx_2^o}{F_{p_2}} = \cdots = \frac{dx_n^o}{F_{p_n}} = d\sigma. \tag{1.46}$$

However, from (1.34), we have at once that

$$dx_i^o = \sum_{r=1}^{n-1} \frac{\partial X_i}{\partial \beta_r} d\beta_r, \tag{1.47}$$

so that in terms of (1.46) and (1.47) the condition $J \neq 0$ is seen to be the condition that $\sum^o$ should not be tangent to the Monge cone. In addition to this, we have already seen that $J \neq 0$ is the condition that it is possible to solve (1.39) and (1.42) for the n functions P_i on $\sum^o$. Consequently if $\sum^o$ is tangent to a Monge cone, so that $J = 0$, then the P_i are indeterminate and so may be prescribed arbitrarily on each side of such a $\sum^o$. Sub-manifolds $\sum^o$ on which $J = 0$ are called characteristic manifolds, and the arbitrariness of p_i across them implies that they are singular surfaces across which the gradient of an integral surface may be (but is not necessarily) discontinuous, as described in Section 1.1.

Knowledge of the (2n + 1) functions $x_1, x_2, \ldots, x_n$, u, $p_1, p_2, \ldots, p_n$ satisfying (1.29), (1.30) and (1.33) amounts to the specification in $\mathbb{R}^n$ at points Q_σ of the ray $\underline{x} = \underline{x}(\sigma)$, of $u = u(\sigma)$ together with the normal $\underline{p} = \underline{p}(\sigma)$ to the local integral surface at Q_σ, and hence to the specification of the local Monge cone. This set of (2n + 1) quantities is called a characteristic strip. The solution to (1.21) is obtained by piecing together all characteristic strips through points Q_o of $\sum^o$.

To relate the ideas of this section to propagating waves it is only necessary to identify u with the time t and $\mathbb{R}^n$ with n-dimensional Euclidean space. It then follows that $u(\underline{x}) = t$, for successive times $t_1 < t_2 < \ldots$, characterises a manifold, or wavefront, at these successive times. The normal to the manifold in $\mathbb{R}^n$ is $\underline{p}$, and the speed of propagation c of a local tangent hyperplane to this manifold along this normal is

$$c = (p_1^2 + p_2^2 + \ldots + p_n^2)^{-\frac{1}{2}}$$

Interpreted in this way, the characteristic equations determine the position, orientation and normal speed of propagation of the local tangent hyperplanes at time $t = u$. They thus provide a natural description of a propagating wavefront.

The complexity of the form of the integral surface to the general equation (1.21) simplifies at once if that equation is quasilinear, for as it is then linear in the p_i it may be written

$$F \equiv a_1(\underline{x},u)p_1 + a_2(\underline{x},u)p_2 + \ldots + a_n(\underline{x},u)p_n + b(\underline{x},u) = 0. \qquad (1.48)$$

Equations (1.29) then become

$$\frac{dx_i}{d\sigma} = a_i(\underline{x},u) \qquad \text{for } i = 1, 2, \ldots, n,$$

and when combined with (1.30) and $F = 0$ they give

$$\frac{du}{d\sigma} = -b(\underline{x},u). \qquad (1.49)$$

The one parameter Monge cone at any point Q_σ on a ray is then determined from these equations by the system

$$\frac{dx_1}{a_1(\underline{x},u)} = \frac{dx_2}{a_2(\underline{x},u)} = \ldots = \frac{dx_n}{a_n(\underline{x},u)} = \frac{du}{-b(\underline{x},u)} = d\sigma, \qquad (1.50)$$

which defines a unique (apart from sign) direction at Q_σ. Thus the Monge cone degenerates into a single line at each point Q_σ of the ray, with the line being directed along the axis of the cone. Hence the task of determining the Monge cones, which in the general case are necessary to determine the integral surface, will in the quasilinear case simplify to finding the rays defining the axes of the degenerate Monge cones. We shall have more to say about this matter in a later section.

There are two other classes of equation that require identification on account of their importance. These are the classes of semilinear and linear partial differential equations, though as we will subsequently be concerned with what will be called genuine nonlinearity we offer only a few remarks about them.

A single partial differential equation, or system, with independent variables $\underline{x} = \underline{x}(x_1, x_2, \ldots, x_n)$ in $\mathbb{R}^n$ which is of order m, and in which the derivatives of order m appear to degree one with coefficients which are functions only of $\underline{x}$, whilst some or all of the remaining terms involving the dependent variables and their derivatives are nonlinear will be said to be semilinear and of order m. A special case occurs when the equation, or system, is of first order. The nonlinearity is then confined to the term, or terms, containing the undifferentiated dependent variable, or variables.

If, in a system of partial differential equations, all the dependent variables and their derivatives appear to degree one and have coefficients that are functions only of $\underline{x}$, then such a system will be said to be linear. An analogous definition exists for a single equation.

It is at once apparent that the analysis of this present section simplifies immediately in each of these classes, since the coefficients involved in the characteristic equations will not involve u.

1.5 The Eikonal Equation

The first order nonlinear partial differential equation

$$(\nabla \phi)^2 - n^2(\underline{x}) = 0, \tag{1.51}$$

in which ∇ is the gradient operator in $\mathbb{R}^n$, is called the eikonal equation. It occupies a special position amongst the various first order nonlinear partial differential equations of mathematical physics and it occurs in a variety of ways [6]. One important way in which it arises is of considerable relevance both to the general subject of these research notes and to the particular topic just discussed in the previous section. We refer here to the role of the eikonal equation in geometrical optics [16] where its solution is essential if an asymptotic solution is to be found to the Helmholtz equation which describes the time reduced wave equation in $\mathbb{R}^3$.

Because of the intrinsic interest of this problem, and the fact that it was geometrical optics that gave rise to the term ray, let us now outline the derivation of the formal asymptotic solution to the Helmholtz equation

$$(\nabla^2 + k^2n^2(\underline{x})) v = 0 \tag{1.52}$$

as $k \to \infty$. Once this has been accomplished we shall then proceed to a brief discussion of the eikonal equation itself.

We will seek a solution to the Helmholtz equation (1.52) in the asymptotic form

$$v(\underline{x},k) = \exp(ik\emptyset(\underline{x}))Z(\underline{x},k), \tag{1.53}$$

in which

$$Z(\underline{x},k) \sim \sum_{s=0}^{\infty} Z_s(\underline{x})(ik)^{-s}. \tag{1.54}$$

Substitution of (1.53) and (1.54) into (1.52) followed by the usual process of equating to zero the coefficient of each successive power of ik gives the following system of partial differential equations,

$$(\nabla\emptyset)^2 - n^2(\underline{x}) = 0, \tag{1.55}$$

$$2(\nabla\emptyset).(\nabla Z_o) + (\nabla^2\emptyset)Z_o = 0, \tag{1.56}$$

and the recursively defined system for $Z_s(\underline{x})$

$$2(\nabla\emptyset).(\nabla Z_s) + (\nabla^2\emptyset)Z_s - \nabla^2 Z_{s-1} = 0, \qquad \text{for } s = 1,2,\ldots . \tag{1.57}$$

The first of these equations is the eikonal equation, whilst the second is known as the transport equation. Once the eikonal equation has been solved, and the result used to obtain $Z_o(\underline{x})$ from the transport equation, then the third equation (1.57) may be used recursively to find the higher order correction terms $Z_s(\underline{x})$. When these are known the asymptotic representation (1.53) can be constructed and the problem is formally solved. The problem we now consider is the solution of (1.51) which is merely a special case of the general first order nonlinear partial differential equation (1.21) in which $F \equiv (\nabla\emptyset)^2 - n^2(\underline{x})$.

The characteristic equations (1.29) (1.30) and (1.31) may be used immediately, but it will be instructive to retain the parameter $k(\underline{x})$ introduced in (1.24) rather than set it equal to unity as in (1.29) and when deriving (1.30) and (1.33). When this is done, the characteristic equations for the eikonal equation (1.51) take the form

$$\frac{d\underline{x}}{d\sigma} = 2k\underline{p}, \tag{1.58}$$

$$\frac{d\emptyset}{d\sigma} = 2kn^2, \tag{1.59}$$

and

$$\frac{d\underline{p}}{d\sigma} = 2kn\nabla n, \tag{1.60}$$

where, since $\underline{p} = \nabla\emptyset$, the eikonal equation (1.51) itself has reduced to

$$p^2 = n^2. \tag{1.61}$$

When written in this form it is easy to see the role played by $k(\underline{x})$ in the connection between σ and arc length along a ray. If σ is to be the actual arc length along a ray then it is necessary to require that $|d\underline{x}/d\sigma| = 1$. That is, we must require that

$$|2k\underline{p}| = 2|k|\underline{p}|| = 2|kn| = 1,$$

from which we conclude that the appropriate choice for k is

$$k = \frac{1}{2n(\underline{x})} \quad . \tag{1.62}$$

Thus in this case the proportionality factor k will, in general depend on $\underline{x}$. For many purposes there is no necessity to adopt such a parameterisation, and when this is the case it is simpler to set $k \equiv 1/2$ so that σ is merely a function of arc length, as was the case previously.

We have already seen that the system (1.58) to (1.60) of first order ordinary differential equations specifies the rays $\underline{x} = \underline{x}(\sigma)$, the corresponding normal $\underline{p} = \underline{p}(\sigma)$ to the surfaces $\emptyset$ = const. and the function $\emptyset(\sigma)$ itself which, in the context of geometrical optics, is called the wave phase. Here, the rays $\underline{x} = \underline{x}(\sigma)$ are the paths followed by the so called light rays in geometrical optics and from (1.58) they are seen to be the orthogonal trajectories to the equi-phase surfaces.

The specification of initial data corresponding to $\sigma = 0$ then enables the evolution of the phase surface to be determined as a function of σ though, as remarked earlier, the initial data for the characteristic equations has to be determined from the data given for the original partial differential (eikonal) equation.

Suppose that there exists an initial surface $\mathcal{S}_o$ in $\mathbb{R}^3$, corresponding to $\sigma = 0$, which is given in the parametric form

$$\underline{x} = \underline{X}_o(\beta_1, \beta_2), \tag{1.63}$$

and assume further that on $\mathcal{S}_o$ the function $\phi(\underline{x})$ is specified by

$$\phi = \Phi_o(\beta_1, \beta_2). \tag{1.64}$$

Then if $\mathcal{S}_o$ is non-characteristic, so that in general $\nabla\phi$ has a non-zero component $\nabla_\nu \phi$ normal to $\mathcal{S}_o$, the arbitrariness of the sign of the normal $\underline{\nu}$ to $\mathcal{S}_o$ implies that for a chosen $\underline{\nu}$, at any point P of $\mathcal{S}_o$, there are two values of $\underline{p}$. These have the same tangential component $\nabla_\tau\phi$ but equal and opposite components along the vector $\underline{\nu}$. Denote the vector $\underline{p}$ at point P of $\mathcal{S}_o$ for which $\underline{p}.\underline{\nu} > 0$ by $\underline{p}^+$ and the vector $\underline{p}$ for which $\underline{p}.\underline{\nu} < 0$ by $\underline{p}^-$. Then $\underline{p}^+$ represents an outgoing vector from $\mathcal{S}_o$ at P. This situation is illustrated in Fig. 2. The sense of $\underline{p}$ may thus be identified with the direction of propagation of the surface or wavefront ϕ = const, and it is conventional to parameterise the surfaces ϕ = const. in such a manner that ϕ increases in the direction of propagation.

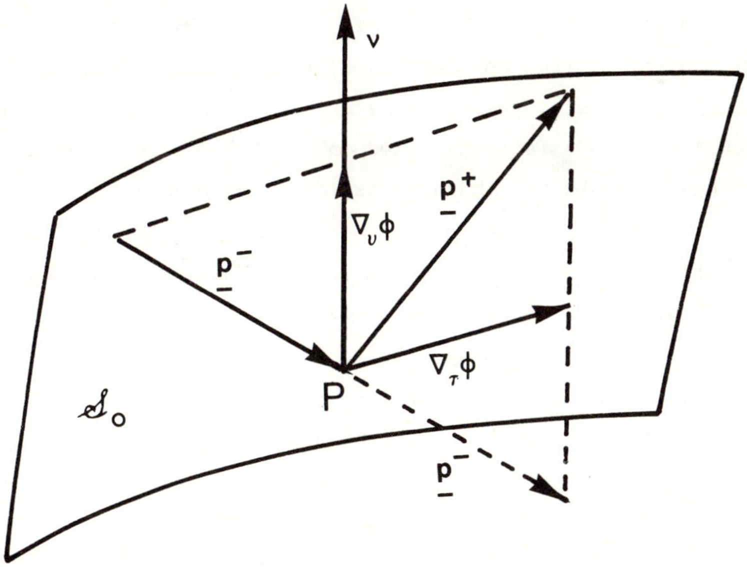

Fig. 2. Incoming and Outgoing Vectors

To complete this brief examination of the eikonal equation, let us examine the consequences of having $n(\underline{x}) = n_o$ = const, and a non-characteristic initial surface $\mathcal{S}_o$ on which data is prescribed for the characteristic equations (1.58) to (1.60) in which, for convenience, we set $k = 1/2$. Equation (1.60) immediately integrates to yield $\underline{p}$ = const. along each individual ray. Thus, as a ray is identified by the ordered pair of parameters (β_1,β_2), we have

$$\underline{p} = \underline{P}_o\,(\beta_1,\beta_2) \tag{1.65}$$

where $\underline{P}_o\,(\beta_1,\beta_2)$ is the constant value of $\underline{p}$ associated with the point corresponding to (β_1,β_2) on $\mathcal{S}_o$. Employing this result in (1.58) with $k = 1/2$ then gives

$$\underline{x} = \underline{X}_o\,(\beta_1,\ \beta_2) + \underline{P}_o\,(\beta_1,\beta_2)\sigma \tag{1.66}$$

The solution $\emptyset$ itself then follows directly from (1.59) with $k = 1/2$ and has the form

$$\emptyset \ = \ \Phi_o(\beta_1,\beta_2) + n_o^{\,2}\sigma. \tag{1.67}$$

Although the parameters σ,β_1 and β_2 may, in theory, be eliminated between equations (1.65) to (1.67) to yield $\emptyset = \emptyset(\underline{x})$, this is often not possible. Indeed it is seldom even desirable to perform this elimination, since the simplicity of the parametric representation usually aids in the interpretation of the rays in geometrical terms. The nature of the solution represented by equations (1.65) to (1.67) is illustrated in Fig.3. in which the initial surface $\mathcal{S}_o$ is taken to be non-characteristic. Thus nowhere are the rays indicated by arrows tangent to $\mathcal{S}_o$. As the evolution of the initial $\emptyset$ = const. surface with σ is in the sense of the vector $\underline{p}$ it follows from (1.67) that $\emptyset$ increases in the direction of propagation of the initial $\emptyset$ = const. surface.

Singularities in the propagating wavefront will occur, for example, whenever rays intersect and also along lines or surfaces which are tangent to the rays. This is the situation with so called caustic curves [6,16] which are of importance in geometrical optics and in diffraction theory in general. The precise way in which these problems are resolved will not be discussed in these research notes, but suffice it to say here that the problem of extending a solution along a caustic is connected with the fact that the caustic lies on a characteristic surface. Thus the characteristic initial value problem must be studied for the resolution of such problems. Clearly singularities of this type will not arise for the situation depicted in Fig. 3, since there the rays diverge with increasing σ.

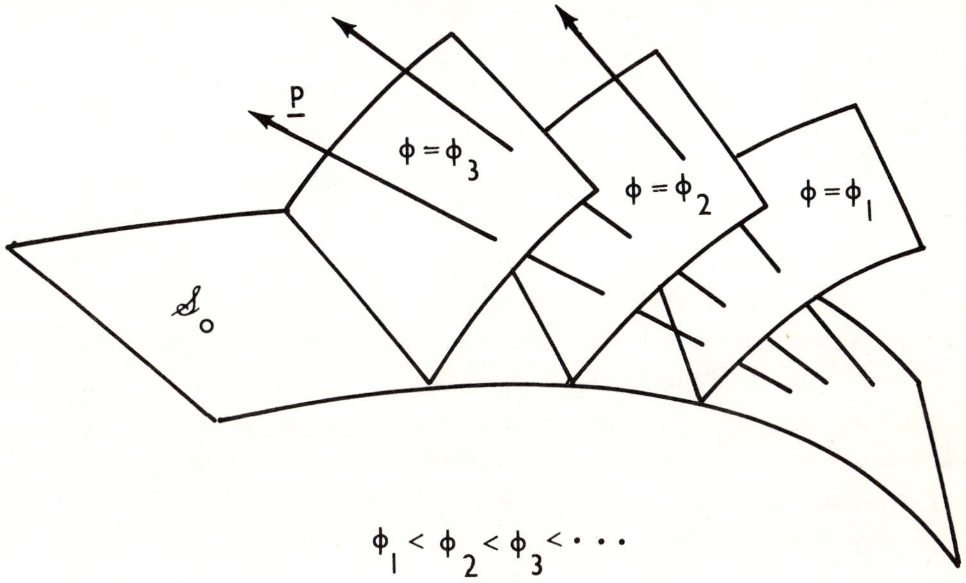

Fig. 3. Non-characteristic initial surface $\mathscr{S}_o$, rays and wavefronts.

1.6 The Scalar Quasilinear First Order Equation with Two Independent Variables

On account of its importance, further reference must be made to the scalar quasilinear equation of first order defined in equation (1.48). In the present section, however, attention will be confined to the case of two independent variables, so that the equation will be defined in $\mathbb{R}^2$. Since the two independent variables concerned will usually be identified with time and a spatial dimension they will be denoted by t and x, respectively, so that $\underline{x} = (t,x)$. Under these conditions equation (1.48) becomes

$$a_1(x,t,u)\frac{\partial u}{\partial t} + a_2(x,t,u)\frac{\partial u}{\partial x} + b(x,t,u) = 0, \qquad (1.68)$$

where we assume a_1, a_2 and b to be continuous functions of their arguments.

The equations determining the degenerate Monge cone that exists in this case follow from (1.50) in the form

$$\frac{dt}{d\sigma} = a_1(x,t,u), \tag{1.69}$$

$$\frac{dx}{d\sigma} = a_2(x,t,u), \tag{1.70}$$

$$\frac{du}{d\sigma} = -b(x,t,u). \tag{1.71}$$

The first pair of equations (1.69) and (1.70) define a family of curves which in $\mathbb{R}^2$ determine the lines to which the Monge cones have degenerated in this special case. The family of curves C so defined is the field of characteristic curves for equation (1.68), and curves belonging to this field are determined here in terms of the parameter σ. The third equation (1.71) then determines the variation of the solution u, as a function of σ, along any particular characteristic belonging to this field. Taken together these equations define the solution u to equation (1.68) at points of the (x, t)-plane in terms of "initial" data assigned for u along some "initial" curve Γ in the (x, t)-plane. This curve Γ then takes the place of the initial sub-manifold $\sum^0$ in the more general theory and it may be defined in terms of a parameter β by writing

$$\Gamma : \quad t = T(\beta) \quad , \quad x = X(\beta), \tag{1.72}$$

where T and X are given functions of β. Similarly, u may be specified along Γ in terms of β by writing

$$u = U(\beta), \tag{1.73}$$

with U a given function of β.

If desired, σ may be eliminated between equations (1.69) to (1.71), when the characteristic field C is defined by

$$C : \quad \frac{dx}{dt} = \frac{a_2(x,t,u)}{a_1(x,t,u)} \quad , \tag{1.74}$$

and u is then defined along a characteristic curve C in terms of t by

$$\frac{du}{dt} = -\frac{b(x,t,u)}{a_1(x,t,u)} \ . \tag{1.75}$$

We remark here that there is a natural direction associated with characteristics of the field C determined by the sense of description of the characteristics as σ, or t, increases.

Although equations (1.69) to (1.71), subject to (1.72) and (1.73), serve to define the solution u, it is only rarely that the precise analytical form of the solution may be found from these equations. This is on account of the nonlinearity which requires u to be known before the characteristic field may be determined.

The choice of the initial curve Γ is not, of course, completely arbitrary, as there are restrictions on it just as there were restrictions on the choice of initial sub-manifold $\sum^{o}$ in the more general problem examined in Section 1.4. The restrictions on Γ are implied by the condition (1.44) which, for equation (1.68), becomes

$$\begin{vmatrix} \frac{dT}{d\beta} & a_1(X,T,U) \\ \frac{dX}{d\beta} & a_2(X,T,U) \end{vmatrix} \neq 0. \tag{1.76}$$

If, now, (1.76) is interpreted as the condition that the two column vectors comprising the determinant should not be proportional, then condition (1.76) asserts that

$$(dX/d\beta)/(dT/d\beta) \neq a_2(X,T,U)/a_1(X,T,U). \tag{1.77}$$

The left hand side of this inequality defines the gradient of Γ, whilst from equation (1.74) we see that the right hand side defines the gradient of the characteristic curve C at that same point. Hence condition (1.76) simply requires that the initial curve Γ should nowhere be tangent to a characteristic. We conclude that for Γ to be a properly defined initial curve it is necessary that it should be non-characteristic.

Although the specification of Cauchy data on a non-characteristic curve Γ will ensure the local existence of a unique differentiable solution it will not ensure its existence in the large. This may be appreciated by examining Fig. 4. in which Γ is a non-characteristic initial curve. Along the arc AB of Γ the characteristic field is diverging with increasing time t. Consequently, as the characteristics do not intersect they transport the solution without ambiguity. However, along the arc BC of Γ the characteristic field is convergent so that, in general, where characteristics intersect the solution will be non-unique. In this case the first occurrence of non-uniqueness can be determined by examining the envelope of the characteristics.

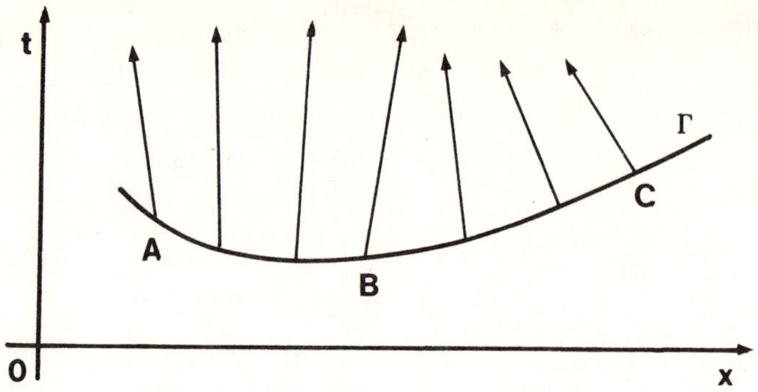

Fig.4. Non-characteristic initial curve Γ. Arrows indicate the direction along characteristics associated with increasing σ or t.

This limitation of the region of existence of a unique differentiable solution to a quasilinear equation is best illustrated by example. Let us consider the equation

$$\frac{\partial u}{\partial t} + f(u) \frac{\partial u}{\partial x} = 0, \tag{1.78}$$

with the initial line Γ chosen to be the line $t = 0$ on which

$$u(x,0) = g(x). \tag{1.79}$$

Then equation (1.74) determining the characteristic field C becomes

$$C : \quad \frac{dx}{dt} = f(u), \tag{1.80}$$

whilst equation (1.75) determining the variation of u along the characteristic curves becomes

$$\frac{du}{dt} = 0. \tag{1.81}$$

As (1.81) implies $u = \text{const.}$ along each characteristic, which are thus straight lines though with different gradients, it follows at once that the characteristic through the point $(s,0)$ on the initial line $t = 0$ has the equation

$$x = s + tf(g(s)), \tag{1.82}$$

while along this characteristic

$$u = g(s). \tag{1.83}$$

Elimination of s between (1.82) and (1.83) then gives as the implicit solution for u

$$u = g(x - tf(u)) . \tag{1.84}$$

When expressed in parametric form the envelope of the characteristics (1.82), with s as parameter, is easily shown to be given by

$$x = s + t\,f(g(s)) \quad , \quad t = -1/f'(g(s))\,g'(s). \tag{1.85}$$

The construction and form of part of the envelope for the special case $f(u) = u$, $g(x) = a \sin x$ with $a > 0$ is shown in Fig.5.

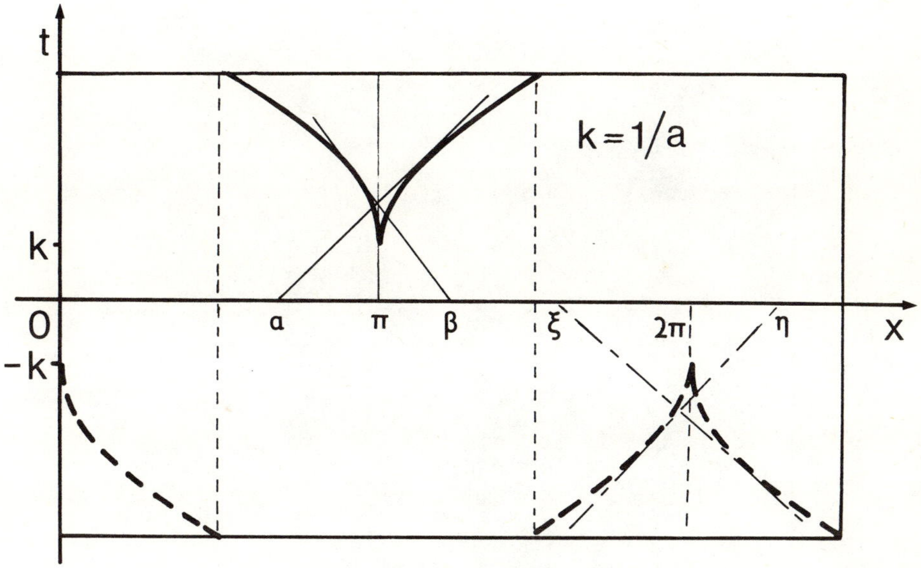

Fig.5. Construction of part of the positive and negative branches of the envelope.

Due to the periodicity of the initial data the envelope itself is periodic in x with period 2π, though to illustrate its construction both for positive and negative t the domain $0 \leq x \leq 5\pi/2$ has been considered. Representative characteristics giving rise to the envelope for $t > 0$ are shown constructed through the points $(\alpha,0)$ and $(\beta,0)$ of the initial line. The corresponding construction giving rise to the envelope for $t < 0$ is shown with characteristics constructed through the points $(\xi,0)$ and $(\eta,0)$ of the initial line. The two envelopes are displaced reflections of one another in the x-axis.

That part of the envelope which has been constructed for $t > 0$ is shown as a heavy full line, while the corresponding envelope for $t < 0$ is shown as a heavy dotted line. The complete region in the domain $0 \leq x \leq 2\pi$ in which a unique solution to (1.78) is determined by the initial data $u(x,0) = a \sin x$ is shown as the shaded region in Fig.6 which is, of course, periodic in x with period 2π. Naturally the solution for $t < 0$ might have no physical meaning if t denotes time. If t is regarded as time, then the first elapsed time at which non-uniqueness occurs for $0 \leq x \leq 2\pi$ is when $t = 1/a$ and $x = \pi$.

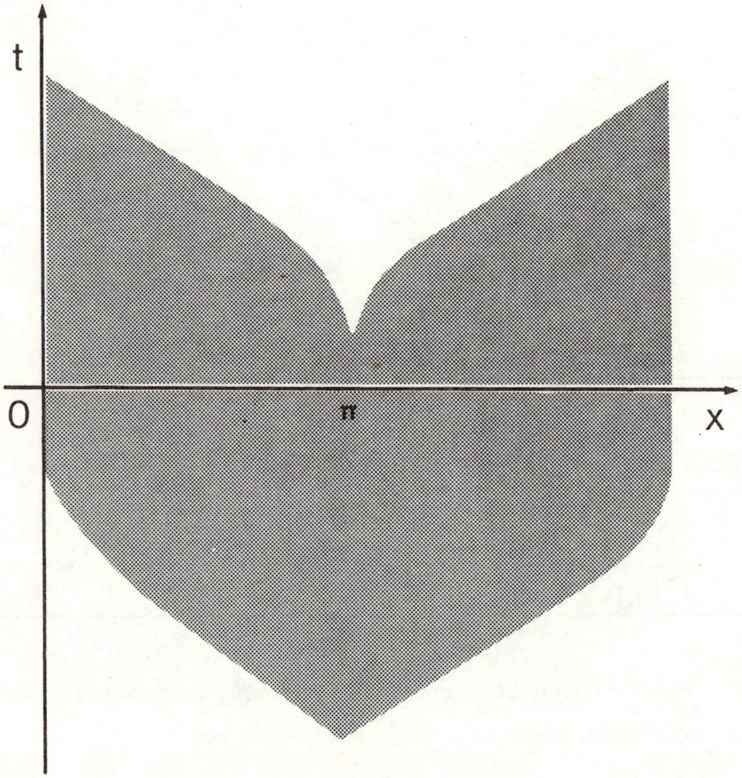

Fig. 6. Region for $0 \leq x \leq 2\pi$ which has a unique solution. The complete region is periodic in x with period 2π.

It is important to remark here that the confining of a unique differentiable solution to such a region in this manner is not due to any lack of continuity or differentiability properties on the part of the coefficients or initial data as represented by the composite function $f(g(s))$. The behaviour of the family of straight lines comprising the characteristic field for

(1.78) which define the envelope, when it exists, is determined solely by the function $f(g(s))$ occurring in (1.82), where $(s,0)$ is any point of the initial line. If this function is monotonic increasing with s, then the characteristics diverge and intersection will not occur for $t > 0$. However, if it is monotonic decreasing, or if it is just an arbitrary function, then intersection of characteristics will occur with $t > 0$ for some x-domains, as in Fig.5. That the non-uniqueness of the solution is not directly related to the continuity and differentiability of $f(g(s))$ can be seen by observing the fact that the characteristics through $(s_1,0)$ and $(s_2,0)$ will intersect for $t > 0$ if $f(g(s_2)) < f(g(s_1))$ with $s_2 > s_1$, without any further restriction being placed on $f(g(s))$. Thus the solution will still be furnished by (1.82) and (1.83) when, for example, u is only Lipschitz continuous on the initial line which is a situation that will concern us later.

If the precise forms of $f(u)$, $g(s)$ are not known then the exact envelope of the characteristics cannot be determined. However, the second equation in (1.85) may still be used to furnish an estimate for the upper limit T to the strip $0 \leq t \leq T$ in which a unique solution of (1.78) exists, subject to (1.79), when all that is known is that $f(u)$ is differentiable and $g'(s)$ is bounded. This follows because if $|g(s)| \leq K$, $|g'(s)| \leq L$ and $\max |f'(\zeta)| = M$ for $|\zeta| \leq K$, then

$$T < t_c = 1/LM.$$

By virtue of the definition of T it also follows that within the strip $|u| \leq K$. The provision of a priori estimates of this type for a solution is frequently extremely valuable, as is the determination of the region in which there exists a unique classical solution. In the case of the example employed in Figs.5 and 6 the above inequality reduces to $T < t_c = 1/a$, which in the case of a time evolution problem represents the elapsed at which the cusp on the envelope first forms.

An alternative approach to the solution of (1.68) that does impose differentiability conditions is to seek a Taylor series solution subject to (1.72) and (1.73). This requires the severe restriction that a_1, a_2, b and U should be analytic. To see how this arises we remark first that as $u = U(\beta)$ is known along Γ, so also is $u_x = k_1(\beta)$, say, and via (1.68) itself, $u_t = k_2(\beta)$. Thus to construct the Taylor series for u the higher derivatives must be constructed from the knowledge of $U(\beta)$, $k_1(\beta)$, $k_2(\beta)$ and (1.68) itself. It will suffice to show how the second order derivatives may be determined, for the others follow in similar fashion.

Along the characteristic field C defined by (1.74) the directional derivative operator d/dt implied by the left hand side of (1.68) is

$$\frac{d}{dt} \equiv \frac{\partial}{\partial t} + \left(\frac{a_2}{a_1}\right)\frac{\partial}{\partial x} .$$

Acting on (1.75) with this operator then gives

$$\frac{d^2u}{dt^2} = \frac{d}{dt}\left(-\frac{b}{a_1}\right)$$

or, retaining only the second order derivatives on the left hand side,

$$\frac{\partial^2 u}{\partial t^2} + \left(\frac{2a_2}{a_1}\right)\frac{\partial^2 u}{\partial x \partial t} + \left(\frac{a_2^2}{a_1^2}\right)\frac{\partial^2 u}{\partial x^2} = h(x,t,u,\frac{\partial u}{\partial t}, \frac{\partial u}{\partial x}) \tag{1.86}$$

along C.

The directional derivative along the initial curve Γ is

$$\frac{d}{dt} \equiv \frac{\partial}{\partial t} + \left(\frac{dX}{d\beta}\right)\left(\frac{d\beta}{dT}\right)\frac{\partial}{\partial x} \equiv \frac{\partial}{\partial t} + \left(\frac{dX}{dT}\right)\frac{\partial}{\partial x},$$

where $T(\beta)$, $X(\beta)$, are defined in (1.72). Hence acting on u_x and u_t with this operator gives the two equations

$$\frac{d}{dt}\left(\frac{\partial u}{\partial t}\right) = \frac{dk_2}{dt}, \quad \frac{d}{dt}\left(\frac{\partial u}{\partial x}\right) = \frac{dk_1}{dt}$$

along Γ. Thus, denoting the respective right hand sides by ℓ, m, we may augment (1.86) by the two equations

$$\frac{\partial^2 u}{\partial t^2} + \left(\frac{dX}{dT}\right)\frac{\partial^2 u}{\partial x \partial t} = \ell(x,t,u,\frac{\partial u}{\partial t}, \frac{\partial u}{\partial x}), \tag{1.87}$$

$$\frac{\partial^2 u}{\partial x \partial t} + \left(\frac{dX}{dT}\right)\frac{\partial^2 u}{\partial x^2} = m(x,t,u,\frac{\partial u}{\partial t}, \frac{\partial u}{\partial x}), \tag{1.88}$$

along Γ. If we consider (1.86) for points of the characteristic field C on Γ we may solve (1.86) to (1.88) for the three second order derivatives at points P on Γ. These can be solved provided the determinant $\Delta \neq 0$, where when evaluated at such a point P we have

$$\Delta_P = \begin{vmatrix} 1 & 2(a_2/a_1) & a_2^2/a_1^2 \\ 1 & (dX/dT) & 0 \\ 0 & 1 & (dX/dT) \end{vmatrix}_P = \left(\frac{dX}{dT} - \frac{a_2}{a_1}\right)_P^2 \qquad (1.89)$$

We conclude that it will not be possible to deduce the second order derivatives if $\Delta_P = 0$, that is, if

$$\left(\frac{dX}{dT}\right)_P = \left(\frac{a_2}{a_1}\right)_P \quad . \qquad (1.90)$$

This is just the condition that Γ is tangent to the characteristic at P, so that once again the non-characteristic nature of Γ is required if a solution is to be obtained for points P on Γ. To obtain the Taylor series all higher order derivatives need to be deduced, so that a_1, a_2, b and U require to be analytic if this process is to be continued indefinitely.

The establishment of the existence of an analytic solution of this form together with the proof that it is convergent and unique within the neighbourhood of Γ constitutes the Cauchy-Kowalewski theorem [4,6]. Since our concern will not be with analytic data or coefficients we shall not pursue this approach further.

1.7 Uniqueness of Solutions

We now prove the following uniqueness theorem for a class of scalar quasilinear first order equations which form an important special case of the class defined in (1.68). Here the coefficients are restricted to being functions of u, but the method of proof employed may be extended to the class (1.68) itself, and also to the quasilinear first order hyperbolic systems that will be considered later.

Theorem 1.1

Let the equation

$$\frac{\partial u}{\partial t} + a(u)\frac{\partial u}{\partial x} + b(u) = 0$$

have coefficients a(u), b(u) that are Lipschitz continuous. Further, let initial data

$$u(x,0) = U(x)$$

be defined on the initial line $t = 0$ which will be assumed to be non-characteristic. Then a sufficient condition that the solution u should be unique in a region $\mathcal{E}$ intersected by the initial line is that $|\partial u/\partial x|$ is bounded in $\mathcal{E}$.

Proof

In an interval J of $\mathbb{R}^1$ let $a(u)$, $b(u)$ satisfy the strong uniformity condition provided by a Lipschitz condition of order 1 with the respective Lipschitz constants K, L, so that

$$|a(u) - a(v)| < K|u-v| \text{ and } |b(u) - b(v)| < L|u-v|$$

for $u, v \in J$. In addition, let $|\partial u/\partial x|$ be bounded in some region $\mathcal{E}$ which is intersected by the initial line, so that we may write

$$\left|\frac{\partial u}{\partial x}\right| < M,$$

with M a constant.

Now suppose that a different solution v exists in $\mathcal{E}$ satisfying the same equation as u, namely

$$\frac{\partial v}{\partial t} + a(v)\frac{\partial v}{\partial x} + b(v) = 0,$$

and also the same initial condition so that

$$u(x,0) = v(x,0) = U(x).$$

Then writing $w = u-v$ and subtracting the equations satisfied by u, v, we find the equation satisfied by w to be

$$\frac{\partial w}{\partial t} + a(u)\frac{\partial w}{\partial x} + (a(u) - a(v))\frac{\partial v}{\partial x} + b(u) - b(v) = 0.$$

Our object will now be to prove that $w \equiv 0$ in $\mathcal{E}$ so that the solutions u and v must be identical in $\mathcal{E}$.

In terms of the characteristic field C associated with w and defined by

$$C : \frac{dx}{dt} = a(u),$$

the equation for w may be written

$$\frac{dw}{dt} = (a(v)-a(u))\frac{\partial v}{\partial x} + b(v) - b(u)$$

along the characteristics C.

Now consider a point $(\xi,0)$ in $\mathcal{E}$ on the initial line and denote by C_ξ the characteristic curve through $(\xi,0)$. Then along C_ξ the solution w will become some function of t, say W(t), so that integrating the ordinary differential equation for w along C_ξ from $t = 0$ to $t = \tau$ gives

$$W(\tau) = W(0) + \int_0^\tau (a(v)-a(u))\frac{\partial v}{\partial x}\, dt + \int_0^\tau (b(u)-b(v))dt,$$

where $w(\tau)$ is evaluated at $t = \tau$ on C_ξ and $W(0) = w(\xi,0)$. The use of elementary inequalities then shows that

$$|W(\tau)| \leq |W(0)| + \int_0^\tau |a(v)-a(u)|\left|\frac{\partial v}{\partial x}\right| dt + \int_0^\tau |b(u)-b(v)|dt \quad \text{along } C_\xi.$$

However $W(0) = 0$, because u, v satisfy the same initial conditions, so that use of the Lipschitz conditions satisfied by the coefficients and of the boundedness condition for $|\partial u/\partial x|$, and hence for $|\partial v/\partial x|$, leads to the result

$$|W(\tau)| \leq (L + KM)\int_0^\tau |u-v|dt,$$

and hence to the final integral inequality

$$|W(\tau)| \leq (L + KM)\int_0^\tau |W(t)|dt \text{ along } C_\xi.$$

Now one form of Gronwall's lemma [17] asserts that if θ, h are continuous non-negative functions in a closed interval $[0,\tau]$, and

$$\theta(\tau) \leq c + \int_0^\tau h(t)\ \theta(t)dt$$

where $c \geq 0$ is a constant,

then

$$\theta(\tau) \leq c \exp\left(\int_0^\tau h(t)dt\right).$$

Making the identifications $\theta \equiv W$, $h \equiv L + KM$ and $c \equiv 0$ gives the immediate result $W(\tau) \equiv 0$ along C_ξ. As $(\xi,0)$ was any point on the initial line in $\mathcal{E}$

we conclude that $w(x,t) \equiv 0$ in $\mathcal{E}$, so that $u \equiv v$, thereby establishing uniqueness. ▌

The existence of the derivative $\partial u/\partial x$ taken together with its boundedness condition imply that $u(x,t)$ is Lipschitz continuous of order 1 with respect to x. In general, this suggests that to determine when a solution ceases to be unique we should consider it to belong to the class of Lipschitz continuous solutions and seek to determine when it ceases to be Lipschitz continuous. This will, indeed, be one form of approach that will be adopted in a subsequent chapter. In terms of general wavefront evolution, as illustrated in Fig.1, this approach may be interpreted as seeking the position on the wavefront trace at which the slope of the disturbance wave surface immediately behind the wavefront becomes unbounded.

1.8 Well-Posed Problems

The notion of a well-posed problem finds its origins in the early work of Hadamard [18]. He considered a problem concerning a partial differential equation, or system, to be well-posed, or correctly set, if the data imposed on the equation, or system, to identify a particular solution is such that,

i) a solution exists,

ii) the solution so determined is unique, and

iii) the solution depends continuously on the data.

In the event that these three conditions are not all satisfied the problem is said to be improperly-posed.

The ideas underlying these requirements for a well-posed problem are simply that in the physical world a solution is usually expected to exist to a real problem and, furthermore, the solution is expected to be unique. However still more is expected of the solution to a physical problem, since when the data used to specify a particular solution is changed slightly then it normally is anticipated that the solution will only exhibit a correspondingly small change. Naturally, when expressed in mathematical terms, this last requirement necessitates the specification of the class to which the solution belongs and the criterion by which the continuous dependence of the solution on the data is to be judged. Examples of simple improperly-posed problems belonging to each of these three categories are to be found below, the first two of which concern scalar first order equations and the third a scalar second order equation. For more information see the works by John [34] and Lavrentiev [35].

(i) Non-Existence of Solution

Consider the linear first order scalar equation

$$\frac{\partial u}{\partial t} + \frac{\partial u}{\partial x} = 0,$$

subject to the two conditions

a) $u(x,0) = x$ and b) $u(0,t) = f(t)$ with $f(0) = 0$.

Then the solution satisfying both the equation and the data in a) is easily seen to be $u(x,t) = x-t$. Consequently $u(0,t) = -t$ showing that the data represented by a) and b) is over-prescribed, so that no solution will exist, unless $f(t) \equiv -t$.

(ii) Non-Uniqueness of Solution

Consider the same equation as in i) above but with conditions a) and b) replaced by $u_x(x,0) = 1$ on the initial line $t=0$. Then the data implies $u(x,0) = x+c$, with c an arbitrary constant of integration. The solution is thus $u(x,t) = x-t+c$ which is not unique because of the arbitrariness of c.

(iii) Non-Continuous Dependence of Solution on Data

The best known example in this category is the following one due to Hadamard and concerns Laplace's equation. Consider the equation

$$\frac{\partial^2 u}{\partial x^2} + \frac{\partial^2 u}{\partial y^2} = 0$$

subject to the Cauchy data

$$u(x,0) = 0, \quad u_y(x,0) = \frac{1}{n} \sin nx.$$

Then the solution is easily found to be

$$u(x,y) = \frac{1}{n^2} \sin nx \sinh ny.$$

However as $n \to \infty$ so the Cauchy data approaches zero uniformly whilst the solution oscillates unboundedly for $y \neq 0$. This behaviour of the solution is unexpected since the solution to the equation with identically zero Cauchy data is $u(x,t) \equiv 0$. This example serves to illustrate that Cauchy data is inappropriate for Laplace's equation. The reason for this behaviour is, of course, because Laplace's equation is an elliptic equation. Furthermore, Cauchy data for this elliptic equation is being imposed on an open region.

2 Hyperbolic systems and characteristics

2.1 Hyperbolicity and First Order Quasilinear Systems

Let us interpret the meaning of hyperbolicity in terms of the first order quasilinear system of equations introduced in equation (1.10). Thus in $\mathbb{R}^n \times t$ we shall work with the properly determined system

$$A_o(U,\underline{x},t)U_t + \sum_{i=1}^{n} A_i(U,\underline{x},t)U_{x_i} + B(U,\underline{x},t) = 0, \tag{2.1}$$

where $U(\underline{x},t)$ is a column vector with the m elements $u_1(\underline{x},t)$, $u_2(\underline{x},t),\ldots,$ $u_m(\underline{x},t)$, $\underline{x} = (x_1,x_2,\ldots,x_n)$ is a vector in $\mathbb{R}^n$, the $A_i(U,\underline{x},t)$ are $m \times m$ matrices with elements dependent on U, $\underline{x}$ and t and $B(U,\underline{x},t)$ is a column vector with elements $b_1(U,\underline{x},t)$, $b_2(U,\underline{x},t),\ldots,b_m(U,\underline{x},t)$ while the suffixes t and x_i denote partial differentiation.

Now the basic idea underlying the hyperbolicity of a system is that the Cauchy problem should be well-posed for it. For the first order system (2.1) the Cauchy problem amounts to specifying U at points on some initial manifold $\mathscr{S}$ in $\mathbb{R}^{n-1} \times t$, so that the system will be hyperbolic when this data is sufficient to determine a unique solution that depends continuously on the data specified at points of $\mathscr{S}$.

With this idea in mind, and in keeping with the geometrical approach to wavefronts that has been adopted so far, let us now seek to determine when it is possible to so group terms of (2.1) that they express the derivative of U normal to $\mathscr{S}$ in terms of derivatives of U in $\mathscr{S}$ and the remaining terms of (2.1). This will then enable the determination of the derivative of U normal to $\mathscr{S}$ (exterior derivative) in terms of the Cauchy data and its derivatives in $\mathscr{S}$ (interior derivatives).

The problem of determining the derivative of U normal to $\mathscr{S}$ is most easily resolved by considering the effect on system (2.1) of a change to new coordinates, one of which is so chosen that the manifold $\mathscr{S}$ is embedded in the family of coordinate manifolds associated with that coordinate. The derivative with respect to that coordinate will then be the only exterior derivative.

Since our objective is to study evolution problems, we shall leave the time variable t unchanged but replace the n component spatial vector $\underline{x} = (x_1, x_2, \ldots, x_n)$ by the new n component vector $\underline{\varphi} = (\varphi_1, \varphi_2, \ldots, \varphi_n)$ where the components $\varphi_i = \varphi_i(\underline{x}, t)$ are assumed to be differentiable functions of their arguments. The manifold $\mathscr{S}$ itself will be taken to be associated with the coordinate φ_k and to have the equation $\varphi_k(\underline{x}, t) = a$(const.) and, aside from this restriction, the other φ_i will be chosen arbitrarily.

The transformation of coordinates that is to be considered thus becomes

$$t' = t \text{ and } \varphi_i(\underline{x}, t) = \text{const.} \quad \text{for } i=1,2,\ldots,n, \tag{2.2}$$

and we shall suppose, at least initially, that the transformation is non-singular in the vicinity of $\mathscr{S}$. Consequently at points P of $\mathscr{S}$ we shall assume that the Jacobian

$$J = \frac{\partial(t, x_1, x_2, \ldots, x_n)}{\partial(t', \varphi_1, \varphi_2, \ldots, \varphi_n)} = \begin{vmatrix} \frac{\partial t}{\partial t'} & \frac{\partial x_1}{\partial t'} & \frac{\partial x_2}{\partial t'} & \cdots & \frac{\partial x_n}{\partial t'} \\ \frac{\partial t}{\partial \varphi_1} & \frac{\partial x_1}{\partial \varphi_1} & \frac{\partial x_2}{\partial \varphi_1} & \cdots & \frac{\partial x_n}{\partial \varphi_1} \\ \cdots & \cdots & \cdots & \cdots & \cdots \\ \frac{\partial t}{\partial \varphi_n} & \frac{\partial x_1}{\partial \varphi_n} & \frac{\partial x_2}{\partial \varphi_n} & \cdots & \frac{\partial x_n}{\partial \varphi_n} \end{vmatrix} \neq 0. \tag{2.3}$$

Now in terms of the new variables defined in (2.2) the operators $\partial/\partial t$ and $\partial/\partial x_i$ appearing in (2.1) become

$$\frac{\partial}{\partial t} \equiv \frac{\partial}{\partial t'} + \sum_{j=1}^{n} \frac{\partial \varphi_j}{\partial t} \frac{\partial}{\partial \varphi_j}$$

and

$$\frac{\partial}{\partial x_i} \equiv \sum_{j=1}^{n} \frac{\partial \varphi_j}{\partial x_i} \frac{\partial}{\partial \varphi_j}, \qquad \text{for } i = 1,2,\ldots,n. \tag{2.4}$$

Hence when re-expressed in terms of $(\underline{\varphi}, t')$ equation (2.1) takes the form

$$A_o(U, \underline{x}, t)\left(\frac{\partial U}{\partial t'} + \sum_{j=1}^{n} \frac{\partial \varphi_j}{\partial t} \frac{\partial U}{\partial \varphi_j}\right) + \sum_{i,j=1}^{n} A_i(U, \underline{x}, t) \frac{\partial \varphi_j}{\partial x_i} \frac{\partial U}{\partial \varphi_j} + B(U, \underline{x}, t) = 0. \tag{2.5}$$

As the derivative of U is to be expressed normal to $\mathcal{S}$, and $\mathcal{S}$ has been embedded in the family of coordinate manifolds $\varphi_k(\underline{x},t) = \text{const.}$, it follows that the required derivative is $\partial U/\partial\varphi_k$. The terms involving $\partial U/\partial\varphi_k$ may thus be separated out from (2.5) and the result written in the simpler form

$$\Lambda\frac{\partial U}{\partial\varphi_k} + R = 0, \tag{2.6}$$

where

$$\Lambda = \left(\frac{\partial\varphi_k}{\partial t}A_o(U,\underline{x},t) + \sum_{i=1}^{n}\frac{\partial\varphi_k}{\partial x_i}A_i(U,\underline{x},t)\right) \tag{2.7}$$

and R is a column vector with its m elements dependent upon $\underline{x}$, t, U and $\partial U/\partial\varphi_i$ with $i \neq k$.

Consequently the derivative $\partial U/\partial\varphi_k$ normal to $\mathcal{S}$ may be determined from (2.6) provided Λ^{-1} exists, which implies the condition

$$\det\Lambda = \left|\frac{\partial\varphi_k}{\partial t}A_o(U,\underline{x},t) + \sum_{i=1}^{n}\frac{\partial\varphi_k}{\partial x_i}A_i(U,\underline{x},t)\right| \neq 0. \tag{2.8}$$

Let us now divide $\det\Lambda$ by $|\nabla_x\varphi_k| \equiv \{\sum_{i=1}^{n}(\partial\varphi_k/\partial x_i)^2\}^{\frac{1}{2}}$ and set

$$-\lambda = \frac{\partial\varphi_k/\partial t}{|\nabla_x\varphi_k|}, \quad \nu_i = \frac{\partial\varphi_k/\partial x_i}{|\nabla_x\varphi_k|} \quad \text{for } i = 1,2,\ldots,n, \tag{2.9}$$

so that the unit vector $\underline{\nu} = (\nu_1,\nu_2,\ldots,\nu_n)$ is then the normalised spatial gradient $\nabla_x\varphi_k$ of φ_k. Then, in place of (2.8), we arrive at an equivalent algebraic condition

$$|-\lambda A_o(P) + \nu_1A_1(P) + \nu_2A_2(P) + \ldots + \nu_nA_n(P)| \neq 0, \tag{2.10}$$

involving the set of numbers $\{-\lambda,\nu_1,\nu_2,\ldots,\nu_n\}$, which must be satisfied at a point P of $\mathcal{S}$ in order that $\partial U/\partial\varphi_k$ should be uniquely determined there by (2.6). Here the notation $A_i(P)$ has been employed to signify the value of $A_i(U,\underline{x},t)$ at point P of the manifold $\mathcal{S}$. Condition (2.10) may be written more compactly still as

$$Q(P;\underline{\nu},\lambda) \neq 0, \tag{2.11}$$

if we set

$$Q(P;\underset{\sim}{\nu},\lambda) \equiv \left|\sum_{i=1}^{n} \nu_i A_i(P) - \lambda A_o(P)\right|. \tag{2.12}$$

The expression $Q(P;\underset{\sim}{\nu},\lambda)$ is simply a homogeneous polynomial of degree m in the quantities $\{-\lambda,\nu_1,\nu_2,\dots,\nu_n\}$. We notice that the normal derivative $\partial U/\partial\varphi_k$ will be indeterminate at any point P of a manifold $\mathcal{S}$ for which

$$Q(P;\underset{\sim}{\nu},\lambda) = 0. \tag{2.13}$$

It is this last condition in conjunction with the definition of Λ in (2.7) that serves to categorise manifolds $\mathcal{S}$ with respect to system (2.1) and which also leads to the classification of system (2.1) itself. If now we regard $\underset{\sim}{\nu}$ as a given unit vector then (2.13) becomes a polynomial equation of degree m in λ, and its coefficients will depend on both the unit vector $\underset{\sim}{\nu}$ and the choice of the point P. The polynomial $Q(P;\underset{\sim}{\nu},\lambda)$ is called the characteristic polynomial of system (2.1) with respect to $\mathcal{S}$, and manifolds $\mathcal{S}$ on which $Q(P;\nu,\lambda) = 0$ are called characteristic manifolds. By analogy, manifolds for which $Q(P;\underset{\sim}{\nu},\lambda) \neq 0$ are said to be non-characteristic.

Now the $m \times m$ matrix Λ defined in (2.7) is determined in terms of the direction ratios of the normal $\mathcal{N}$ to the manifold $\mathcal{S}$ at P or, equivalently, in terms of the number λ and the spatial unit vector $\underset{\sim}{\nu}$. The direction $\mathcal{N}$ at a point P of $\mathcal{S}$ is called time-like if the matrix Λ is positive definite, and when this is so the manifold $\mathcal{S}$ itself is said to be space-like at P. In the event that the direction $\mathcal{N}$ is such that the matrix Λ is indefinite, then the direction $\mathcal{N}$ is said to be space-like and the manifold $\mathcal{S}$ itself is said to be time-like at P. We are now in a position to formulate a rigorous definition of hyperbolicity in a manner which is sufficiently general for what is to follow.

Definition 2.1 (Hyperbolicity)

The first order quasilinear system (2.1) will be said to be strictly hyperbolic in the t-direction at P if the zeros $\lambda^{(1)},\lambda^{(2)},\dots,\lambda^{(m)}$ of the characteristic polynomial $Q(P;\underset{\sim}{\nu};\lambda)$ are all real and distinct for all choices of the unit vector $\underset{\sim}{\nu}$ and if the right eigenvectors $r^{(1)},r^{(2)},\dots,r^{(m)}$ satisfying

$$\sum_{i=1}^{n}\left\{\nu_i A_i(P) - \lambda^{(j)} A_o(P)\right\} r^{(j)} = 0 \tag{2.14}$$

span the space E^m occupied by the m element eigenvectors; that is to say if they comprise a set of m linearly independent vectors in the space E^m. The system (2.1) will merely be said to be hyperbolic in the t-direction if the eigenvectors span the space E^m but the eigenvalues, although all real, are not all distinct. ∎

It should be emphasised that this classification is pointwise and that since the system is quasilinear then, in general, the classification will depend not only on the point P but also on the Cauchy data through the solution vector U at P.

This formulation of the definition of hyperbolicity is such that because of the arbitrariness of the unit vector $\underline{\nu}$ the matrix

$$\sum_{i=1}^{n} \nu_i A_i(P) - \lambda A_o(P), \tag{2.15}$$

and hence the matrix Λ to which it is proportional, is always positive definite so that the t-direction is time-like. It is for this reason that the definition makes reference to hyperbolicity in the t-direction. The zeros $\lambda^{(1)},\lambda^{(2)},\ldots,\lambda^{(m)}$ of the characteristic polynomial $Q(P;\underline{\nu};\lambda)$ are, of course, the eigenvalues with respect to $A_o(P)$ of the matrix

$$\sum_{i=1}^{n} \nu_i A_i(P), \tag{2.16}$$

just as the column vectors $r^{(1)},r^{(2)},\ldots,r^{(m)}$ are the corresponding eigenvectors.

It has thus been established that whenever $\mathcal{S}$ is a non-characteristic manifold the Cauchy problem will lead to the determination of the derivative $\partial U/\partial\varphi_k$ normal to $\mathcal{S}$ in the hyperbolic case of system (2.1). This is clearly a necessary first step to the determination of a solution to the Cauchy problem for system (2.1). If the coefficient matrices and Cauchy data are analytic a Taylor series solution could be developed for U just as described in Section 1.6 for a scalar equation. Consequently in this case, at least in the neighbourhood of $\mathcal{S}$, a method of solution to the Cauchy problem for system (2.1) may be arrived at.

The details of the justification of this process, even in the linear case, lie outside the scope of these notes, but even so the proof of existence of a solution under these conditions is too restrictive for our purposes. We shall, henceforth, assume the Cauchy problem to be well-posed for hyperbolic systems and refer to the literature for further discussion of this matter. At an expository level accounts are to be found in the work by Courant and Hilbert [6], Hadamard [18], Mikhlin [4], Hellwig [19] and Garabedian [20]. For a more abstract account we refer to the work of Schauder [21], Leray [22] and Lax [23], while for a treatment of special problems in the large and in the real domain without appeal to analyticity mention should be made of the work of Gårding [24, 25] and Ingersoll [26]. See also Mizohata [2].

It may happen that the $\lambda^{(i)}$ which are the zeros of the characteristic polynomial $Q(P;\underline{\nu},\lambda)$ are complex. This implies that there are no real characteristic manifolds at P for system (2.1), and in such a case the system (2.1) is said to be elliptic in type at P. Intermediate between the hyperbolic and elliptic classifications for system (2.1) comes the case of an ultrahyperbolic system. This occurs when the characteristic polynomial has both real and imaginary zeros. Finally we remark that if for any manifold $\mathcal{S}$ determining a direction $\mathcal{N}$ in $\mathbb{R}^{n-1} \times t$ the quadratic form that is associated with Λ in (2.7) is singular, then system (2.1) is said to be parabolic in that direction. As with the hyperbolic case, when system (2.1) is quasilinear, these classifications are pointwise and will also depend, via the solution U, on the Cauchy data.

It follows at once from the definition of the characteristic polynomial that when the manifold $\mathcal{S}$ is characteristic the normal derivative $\partial U/\partial\varphi_k$ will be indeterminate on it. Consequently under these circumstances the normal derivative $\partial U/\partial\varphi_k$ may be different on opposite sides of a characteristic manifold $\mathcal{S}$ which then provides a direct generalisation of the type of wavefront already described in Section 1.1. A characteristic manifold will automatically have associated with it one of the zeros of the characteristic polynomial $Q(P;\underline{\nu};\lambda)$, say $\lambda = \lambda^{(i)}$, and it then follows at once from (2.14) that the normal derivative $\partial U/\partial\varphi_k$, although not uniquely determined, must be proportional to the corresponding right eigenvector $r^{(i)}$. This important result is one which we will frequently have occasion to use.

When system (2.1) is hyperbolic in the t-direction the coefficient matrix $A_o(U,\underline{x},t)$ will be positive definite, and so will have an inverse $A_o^{-1}(U,\underline{x},t)$. Pre-multiplication of system (2.1) by $A_o^{-1}(U,\underline{x},t)$ will then yield a system in which the coefficient matrix of the vector U_t is the unit matrix I. This is the form in which many hyperbolic systems of interest in the mathematical sciences most naturally occur, and so it is the one with which we shall most frequently work. However, if $A_o(U,\underline{x},t)$ is not symmetric, pre-multiplication of system (2.1) by $A_o^{-1}(U,\underline{x},t)$ will destroy any symmetry that may be possessed by the coefficient matrices $A_i(U,\underline{x},t)$ which it might be desirable to preserve. When A_o is positive definite but depends only on $\underline{x}$ and t this problem may be overcome and such symmetry preserved, while still achieving a unit initial coefficient matrix, if the dependent variable vector is changed.

Specifically, as $A_o(\underline{x},t)$ is assumed to be positive definite, we may always write

$$A_o(\underline{x},t) = M'M, \tag{2.17}$$

where $M(\underline{x},t)$ is a non-singular matrix and the prime denotes the matrix transpose operation. So, defining a new dependent variable vector V in terms of M by the transformation $V = MU$, we find after substitution into (2.1) and pre-multiplication by $(M')^{-1}$ that V satisfies the system

$$V_t + \sum_{i=1}^{n}(M')^{-1}A_i(U,\underline{x},t)M^{-1}V_{x_i} + R = 0. \tag{2.18}$$

Here vector R depends on the vectors V and B and the $m \times m$ matrix $M(\underline{x},t)$ together with its derivatives with respect to t and x_i. Since the matrix $(M')^{-1}A_iM^{-1} = (M^{-1})'\,A_iM^{-1}$, it follows that the new coefficient matrices $(M^{-1})'\,A_iM^{-1}$ for the derivatives V_{x_i} will be symmetric whenever the A_i are symmetric.

This preservation of symmetry is important in the hyperbolic case when only the matrices $A_1, A_2, \ldots, A_n$ are symmetric, for it enables system (2.1) in which $A_o = A_o(\underline{x},t)$ to be transformed into system (2.18) which is then a symmetric hyperbolic system of quasilinear type.

When the notion of hyperbolicity is combined with the fact that the eigenvalues of a symmetric matrix are real we may conclude that

(i) if in the system

$$A_o(U,\underline{x},t)U_t + \sum_{i=1}^{n}A_i(U,\underline{x},t)U_{x_i} + B(U,\underline{x},t) = 0$$

the coefficient matrix $A_o(U,\underline{x},t)$ is symmetric and positive definite and the matrices $A_1(U,\underline{x},t)$, $A_2(U,\underline{x},t), \ldots, A_n(U,\underline{x},t)$ are symmetric, then the system is symmetric hyperbolic, and

(ii) if all the coefficient matrices in the above system are symmetric, then a necessary and sufficient condition that the system should be symmetric hyperbolic is that $A_o(U,\underline{x},t)$ should be positive definite.

From amongst the useful properties of symmetric hyperbolic systems which are of importance, we only single out two for comment here. The first is that if system (2.1) is symmetric hyperbolic in the t-direction, at a point P, then it is also symmetric hyperbolic in any direction $\mathcal{T}$ through point P in $\mathbb{R}^{n-1}_x \times t$ provided $\mathcal{T}$ is contained in a suitably narrow cone $\mathcal{G}$ with vertex at P and axis parallel to the t-direction. The second property is that the uniqueness of solution to the initial value problem may be readily established for symmetric hyperbolic systems.

For a detailed general development of the theory of linear symmetric hyperbolic systems we refer to the work of Friedrichs [27, 28]. Other accounts are to be found in the books by Courant and Hilbert [6], Garabedian [20] and Hellwig [19]. An application of the properties of symmetric hyperbolic systems to geometrical acoustics has been described by Jeffrey [29] and to weak hydromagnetic discontinuities by Bazer and Fleischman [30] and by Bazer and Ericson [31].

2.2 The Classification of Some Special Systems

The ideas contained in the previous section may best be illustrated by application to some typical systems of interest. Let us take for the first illustration the equations of gas dynamics as represented by the first order quasilinear system in Example 1 of section 1.3.

1. Unsteady Isentropic Compressible Gas Flow

The first order quasilinear system (1.14a) already has a unit coefficient matrix for the vector U_t so that the matrix $A_o(U,\underline{x},t) \equiv I$. Combination of the results of (1.14b) and (2.12) then shows that the characteristic polynomial $Q(P;\underline{\nu};\lambda)$ is

$$Q(P;\underline{\nu},\lambda) = \begin{vmatrix} \underline{\nu}\cdot\underline{q} - \lambda & \nu_1\rho & \nu_2\rho & \nu_3\rho \\ \nu_1 a^2/\rho & \underline{\nu}\cdot\underline{q} - \lambda & 0 & 0 \\ \nu_2 a^2/\rho & 0 & \underline{\nu}\cdot\underline{q} - \lambda & 0 \\ \nu_3 a^2/\rho & 0 & 0 & \underline{\nu}\cdot\underline{q} - \lambda \end{vmatrix},$$

where the scalar product $\underline{\nu}.\underline{q} = \nu_1 u + \nu_2 v + \nu_3 w$. Expanding this result and using the fact that $\underline{\nu}$ is a unit vector gives

$$Q(P;\underline{\nu};\lambda) = (\underline{\nu}.\underline{q} - \lambda)^2\{(\underline{\nu}.\underline{q} - \lambda)^2 - a^2\} . \tag{2.19}$$

Hence the zeros $\lambda^{(i)}$ of $Q(P;\underline{\nu};\lambda)$, or equivalently the eigenvalues (with respect to the unit matrix)of the matrix

$$\sum_{i=1}^{3} \nu_i A_i(P) = \begin{bmatrix} \underline{\nu}.\underline{q} & \nu_1\rho & \nu_2\rho & \nu_3\rho \\ \nu_1 a^2/\rho & \underline{\nu}.\underline{q} & 0 & 0 \\ \nu_2 a^2/\rho & 0 & \underline{\nu}.\underline{q} & 0 \\ \nu_3 a^2/\rho & 0 & 0 & \underline{\nu}.\underline{q} \end{bmatrix},$$

are

$$\lambda^{(1)} = \underline{\nu}.\underline{q}+a,\ \lambda^{(2)} = \underline{\nu}.\underline{q}-a,\ \lambda^{(3)} = \underline{\nu}.\underline{q},\ \lambda^{(4)} = \underline{\nu}.\underline{q} \tag{2.20}$$

All four of these eigenvalues are real, though there are only three distinct eigenvalues owing to the factor $(\underline{\nu}.q - \lambda)^2$ in $Q(P;\underline{\nu},\lambda)$.

The eigenvectors corresponding to $\lambda = \lambda^{(1)}$ and $\lambda = \lambda^{(2)}$ that result from solving the equations corresponding to (2.14) are easily shown to be

$$r^{(1)} = \begin{bmatrix} 1 \\ \nu_1 a/\rho \\ \nu_2 a/\rho \\ \nu_3 a/\rho \end{bmatrix}, \qquad r^{(2)} = \begin{bmatrix} 1 \\ -\nu_1 a/\rho \\ -\nu_2 a/\rho \\ -\nu_3 a/\rho \end{bmatrix}, \tag{2.21a}$$

where the first element of each vector has been normalised to unity. Inserting either $\lambda = \lambda^{(3)}$ or $\lambda = \lambda^{(4)}$ into the equations corresponding to (2.14) shows that the elements r_1, r_2, r_3 and r_4 of each of the corresponding right eigenvectors $r^{(3)}$ and $r^{(4)}$ must be such that $r_1 = 0$, while

$$\nu_1 r_2 + \nu_2 r_3 + \nu_3 r_4 = 0.$$

This last result is simply the scalar product $\underline{\nu}.\underline{d} = 0$ involving the unit normal $\underline{\nu}$ and a space vector $\underline{d} = (r_2,r_3,r_4)$. Hence any vector $\underline{d}$ normal to $\underline{\nu}$ will provide a solution, and as we may always find two such linearly independent vectors $\underline{d}' = (r_2',r_3',r_4')$ and $\underline{d}'' = (r_2'',r_3'',r_4'')$ in the plane normal to $\underset{\sim}{\nu}$ we may associate these with $\lambda^{(3)}$ and $\lambda^{(4)}$. The remaining two eigenvectors are thus

$$r^{(3)} = \begin{bmatrix} 0 \\ r_2' \\ r_3' \\ r_4' \end{bmatrix}, \qquad r^{(4)} = \begin{bmatrix} 0 \\ r_2'' \\ r_3'' \\ r_4'' \end{bmatrix} \tag{2.21b}$$

As the eigenvalues in (2.20) are all real, but not all distinct, and the right eigenvectors in (2.21 a,b) span the space E^4, we conclude that the associated system of equations is merely hyperbolic in the t-direction. In fact the system is symmetric hyperbolic as may be seen by pre-multiplying system (1.14a,b) by

$$K = \begin{bmatrix} 1 & 0 & 0 & 0 \\ 0 & \rho^2/a^2 & 0 & 0 \\ 0 & 0 & \rho^2/a^2 & 0 \\ 0 & 0 & 0 & \rho^2/a^2 \end{bmatrix}.$$

The system (1.14) is typical of many that arise in continuum mechanics in that the dependence of the coefficient matrices on $\underline{x}$ and t is only implicit through the elements of the dependent variable vector U.

2. Steady Two-Dimensional Irrotational Isentropic Flow

In fluid mechanics the equations of steady two-dimensional irrotational isentropic flow are [9]

$$(c^2-u^2)\frac{\partial u}{\partial x} - uv\left(\frac{\partial u}{\partial y} + \frac{\partial v}{\partial x}\right) + (c^2-v^2)\frac{\partial v}{\partial y} = 0,$$

$$\frac{\partial u}{\partial y} - \frac{\partial v}{\partial x} = 0, \tag{2.22}$$

where $\underline{q} = (u,v)$ is the steady state fluid velocity vector and c is the local sound speed. If $c^2-v^2 \neq 0$ these two equations may be written in the matrix form

$$U_y + AU_x = 0, \tag{2.23a}$$

where the matrices U and A are

$$U = \begin{bmatrix} u \\ v \end{bmatrix}, \qquad A = \begin{bmatrix} 0 & -1 \\ \dfrac{c^2-u^2}{c^2-v^2} & \dfrac{-2uv}{(c^2-v^2)} \end{bmatrix}. \tag{2.23b}$$

As this is a two-dimensional problem with y in a time-like role the unit normal $\underset{\sim}{\nu}$ occurring in the characteristic polynomial will simply be the unit vector in the x-direction so that $|\underset{\sim}{\nu}| = \nu_1 = 1$. The characteristic polynomial $Q(P;\underset{\sim}{\nu};\lambda)$ in (2.12) will thus reduce to

$$Q(P;\underset{\sim}{\nu};\lambda) = |A - \lambda I|,$$

and a simple calculation establishes that the two zeros $\lambda^{(\pm)}$ given by $|A - \lambda I| = 0$ are

$$\lambda^{(\pm)} = \{-uv \pm c(u^2+v^2-c^2)^{\frac{1}{2}}\}/(c^2-v^2).$$

These zeros which are, of course, also the eigenvalues of A will be real and distinct provided $u^2+v^2 > c^2$, but they will be complex if $u^2+v^2 < c^2$. When $u^2+v^2 > c^2$ the eigenvectors corresponding to $\lambda^{(\pm)}$ span E^2 and so the system is then strictly hyperbolic in the y-direction. The system is elliptic when $u^2+v^2 < c^2$. As the fluid speed $|\underset{\sim}{q}| = (u^2+v^2)^{\frac{1}{2}}$ it follows that the hyperbolic case corresponds to supersonic flow because $|\underset{\sim}{q}| > c$, whilst the elliptic case corresponds to subsonic flow because $|\underset{\sim}{q}| < c$.

Equations (2.22) could also have been formulated with x in a time-like role provided $c^2-u^2 \neq 0$, when the same conclusions would have been reached concerning the classification. In this application the classification is pointwise and depends only on the fluid speed $|\underset{\sim}{q}|$.

3. Tricomi's Equation

The equation of transonic flow due to Tricomi [6,20,32,33] is the linear variable coefficient equation

$$\frac{\partial^2 u}{\partial x^2} + x\frac{\partial^2 u}{\partial y^2} = 0. \tag{2.24a}$$

Setting $\partial u/\partial x = v$, $\partial u/\partial y = w$ and using the condition for the equality of mixed derivatives $\partial w/\partial x = \partial v/\partial y$ (2.24a) may be replaced by the equivalent system

$$U_x + AU_y = 0, \tag{2.24b}$$

with

$$U = \begin{bmatrix} v \\ w \end{bmatrix}, \quad A = \begin{bmatrix} 0 & x \\ -1 & 0 \end{bmatrix}. \tag{2.24c}$$

Proceeding as in the previous illustration we find that the zeros $\lambda^{(\pm)}$ of the characteristic polynomial are determined by the condition $|A - \lambda I| = 0$, which reduces to the simple condition $\lambda^2 + x = 0$. It is apparent that the classification now depends on x alone. When $x < 0$ the zeros $\lambda^{(\pm)}$ are real and Tricomi's equation is hyperbolic with x time-like, but when $x > 0$ the zeros are complex so that the equation is elliptic. In the event that $x = 0$, so that the characteristic polynomial has a double zero, there is only one eigenvector

$$r = \begin{bmatrix} 0 \\ k \end{bmatrix}, \quad \text{with } k \neq 0 \text{ arbitrary,}$$

satisfying the equation corresponding to (2.14). Thus Tricomi's equation is parabolic along the line $x = 0$. Hence in this case the classification depends only on position and may be summarised as follows:

Tricomi's equation $\dfrac{\partial^2 u}{\partial x^2} + x\dfrac{\partial^2 u}{\partial y^2} = 0$ is

Hyperbolic for $x < 0$;

Parabolic for $x = 0$;

Elliptic for $x > 0$.

4. Shallow Water Wave Approximation

The system of equations in this case has been given in (1.17a,b). The characteristic polynomial has zeros given by $|A - \lambda I| = 0$ which gives rise to the equation $(u-\lambda)^2 - c^2 = 0$ with the two real roots $\lambda^{(\pm)} = u \pm c$. The corresponding eigenvalues $r^{(\pm)}$ of A which satisfy (2.14) span E^2 and are

$$r^{(\pm)} = \begin{bmatrix} 1 \\ \pm\frac{1}{2} \end{bmatrix},$$

so that the system is totally hyperbolic. The fact that the eigenvectors are constant will prove useful in subsequent applications.

5. The Lundquist Equations of Magnetohydrodynamics

The one-dimensional Lundquist equations [31] of compressible magnetohydrodynamic flow are

$$\frac{\partial \mathbb{H}}{\partial t} - \nabla \times [\underline{q} \times \mathbb{H}] = 0 \tag{2.25a}$$

$$\left(\frac{\partial}{\partial t} + \underline{q}.\nabla\right)\rho + \rho\nabla.\underline{q} = 0 \tag{2.25b}$$

$$\left(\frac{\partial}{\partial t} + \underline{q}.\nabla\right)\underline{q} + \frac{1}{\rho}\nabla p(\rho,S) - \frac{\mu}{4\pi\rho}[\nabla \times \mathbb{H}] \times \mathbb{H} = 0 \tag{2.25c}$$

$$\left(\frac{\partial}{\partial t} + \underline{q}.\nabla\right)S = 0 \tag{2.25d}$$

$$\nabla.\mathbb{H} = 0. \tag{2.25e}$$

Here $\mathbb{H} = (H_1, H_2, H_3)$ is the magnetic vector, $\underline{q} = (u,v,w)$ is the fluid velocit[y] ρ is the fluid density, S is the entropy, $p(\rho,S)$ is the fluid pressure and μ is the permeability of the medium. The sound speed a in the medium is relate[d] to $p(\rho,S)$ by the result $a^2 = \partial p/\partial\rho$. Equation (2.25e) may be regarded as a restriction to be imposed on the initial conditions because, by (2.25a), if this condition holds initially then it is true for all time. Expressed in matrix form the seven scalar equations represented by (2.25a) to (2.25d) become

$$U_t + AU_x = 0 \tag{2.26a}$$

where

$$U = [\rho,u,v,H_2,w,H_3,S]' \quad \text{with a prime denoting a transpose} \tag{2.26b}$$

and

$$A = \begin{bmatrix} u & \rho & 0 & 0 & 0 & 0 & 0 \\ a^2/\rho & u & 0 & \mu H_2/4\pi\rho & 0 & \mu H_3/4\pi\rho & p_S/\rho \\ 0 & 0 & u & -\mu H_1/4\pi\rho & 0 & 0 & 0 \\ 0 & H_2 & -H_1 & u & 0 & 0 & 0 \\ 0 & 0 & 0 & 0 & u & -\mu H_1/4\pi\rho & 0 \\ 0 & H_2 & 0 & 0 & -H_1 & u & 0 \\ 0 & 0 & 0 & 0 & 0 & 0 & u \end{bmatrix}. \tag{2.26c}$$

The characteristic polynomial $Q(P;\underline{\nu};\lambda) = |A - \lambda I|$ has seven distinct zeros

$$\begin{aligned} \lambda_f^{(\pm)} &= u \pm c_f, \\ \lambda_s^{(\pm)} &= u \pm c_s, \\ \lambda_t^{(\pm)} &= u \pm b_1, \\ \lambda_e &= u, \end{aligned} \tag{2.27a}$$

with

$$\begin{aligned} b &= \left(\frac{\mu \, \mathbb{H}^2}{4\pi\rho}\right)^{\frac{1}{2}}, \\ b_i &= \left(\frac{\mu H_i^2}{4\pi\rho}\right)^{\frac{1}{2}} \qquad i = 1,2,3, \\ c_f &= \left[\tfrac{1}{2}\left\{(a^2+b^2) + \{(a^2+b^2)^2 - 4a^2 b_1^2\}^{\frac{1}{2}}\right\}\right]^{\frac{1}{2}}, \\ c_s &= \left[\tfrac{1}{2}\left\{(a^2+b^2) - \{(a^2+b^2)^2 - 4a^2 b_1^2\}^{\frac{1}{2}}\right\}\right]^{\frac{1}{2}}. \end{aligned} \tag{2.27b}$$

The corresponding right eigenvectors of A satisfying (2.14) when $H_1 \neq 0$ span E^7 and are

$$r_f^{(\pm)} = \begin{bmatrix} \rho \\ \pm c_f \\ \mp b_1 b_2 c_f/(c_f^2 - b_1^2) \\ H_2 c_f^2/(c_f^2 - b_1^2) \\ \mp b_1 b_3 c_f/(c_f^2 - b_1^2) \\ H_3 c_f^2/(c_f^2 - b_1^2) \\ 0 \end{bmatrix}, \quad r_t^{(\pm)} = \begin{bmatrix} 0 \\ 0 \\ \mp \mathrm{sgn}(H_1)(\underline{\nu} \times \underline{b})_2 \\ (\underline{\nu} \times \mathbb{H})_2 \\ \mp \mathrm{sgn}(H_1)(\underline{\nu} \times \underline{b})_3 \\ (\underline{\nu} \times \mathbb{H})_3 \\ 0 \end{bmatrix} \tag{2.27c}$$

where, as usual, $\underline{\nu}$ is the unit vector in the x-direction and $\underline{b} = (b_1, b_2, b_3)$, with the notations $(\underline{\nu} \times \underline{b})_2$ and $(\underline{\nu} \times \mathbb{H})_3$ denoting, respectively, the second and third components of vectors $\underline{\nu} \times \underline{b}$ and $\underline{\nu} \times \mathbb{H}$, and

$$r_s^{(\pm)} = \begin{bmatrix} \rho \\ \pm c_s \\ \mp b_1 b_2 c_s/(c_s^2 - b_1^2) \\ H_2 c_s^2/(c_3^2 - b_1^2) \\ \mp b_1 b_3 c_s/(c_s^2 - b_1^2) \\ H_3 c_s^2/(c_f^2 - b_1^2) \\ 0 \end{bmatrix}, \quad r_e = \begin{bmatrix} -p_s/a^2 \\ 0 \\ 0 \\ 0 \\ 0 \\ 0 \\ 1 \end{bmatrix}. \tag{2.27d}$$

Here the suffixes f, s, t and e are used, respectively, to identify quantities associated with the fast, slow, transverse and entropy waves that characterise wave propagation in magnetohydrodynamics. The names fast and slow derive from the fact that $c_f > c_s$, because later we shall identify the eigenvalues λ with wave propagation speeds.

6. A Non-Physical Example

The system of equations described by (1.19a,b) provides an interesting example of a non-physical type. The matrix $A(\mu, U_\mu)$ and the solution U_μ depend continuously on a parameter μ, but the zeros $\lambda^{(1)}$, $\lambda^{(2)}$ and $\lambda^{(3)}$ of the characteristic polynomial given by $|A(\mu, U_\mu) - \lambda I| = 0$ are simply

$$\lambda^{(1)} = -1, \quad \lambda^{(2)} = 0 \text{ and } \lambda^{(3)} = 1,$$

which are real and distinct and not only independent of μ, but are also constant. The eigenvectors $r^{(i)}$ corresponding to these $\lambda^{(i)}$ are

$$r^{(1)} = \begin{bmatrix} \cosh \mu u_{2\mu} \\ -1 \\ -\sinh \mu u_{2\mu} \end{bmatrix}, \quad r^{(2)} = \begin{bmatrix} 0 \\ 1 \\ 0 \end{bmatrix}, \quad r^{(3)} = \begin{bmatrix} -\sinh \mu u_{2\mu} \\ 0 \\ \cosh \mu u_{2\mu} \end{bmatrix}, \tag{2.28}$$

which span E^3 showing that the system (1.19a,b) is strictly hyperbolic.

It is a straightfoward matter to see that $r^{(1)}$, $r^{(2)}$ and $r^{(3)}$ will always span E^3 irrespective of the choice of μ and the vector U_μ satisfying (1.19a,b), because det $|r^{(1)}, r^{(2)}, r^{(3)}| \equiv 1$.

2.3 Invariance of Characteristic Manifolds Under a Change of Coordinates

The method of classification based on the Cauchy problem that is described in Section 2.1, and applied in Section 2.2, would not be of value were it to depend on the choice of coordinates used to re-express system (2.1). That it is invariant under a change of coordinate manifolds is really implied by the arbitrariness with which the new coordinate system $\underline{\varphi}$ was chosen together with the notion of an exterior derivative that was used. However the result may be established formally by appeal to a property of Jacobian matrices which derives from the chain rule for differentiation.

To prove this result let us suppose in (2.5) that a further transformation of coordinates is made by setting

$$t'' = t' \text{ and } \Psi_j(\underline{\varphi}, t') = \text{const. for } j = 1,2,\ldots,n, \tag{2.29}$$

where again because we are concerned with evolution problems the time variable has been left unchanged. Then if embedded in the Ψ_k coordinate manifold is the same initial manifold $\mathcal{S}$ in $\mathbb{R}^{n-1} \times t$ (corresponding, say, to $\Psi_k(\underline{\varphi}, t') = b$ (const.)), the normal derivative of U with respect to $\mathcal{S}$ will become $\partial U/\partial \Psi_k$ and it is unique apart from its sign.

Making the further variable change in (2.5) and denoting the coefficient matrix for $\partial U/\partial \Psi_k$ by $\mathcal{G}$ gives, after some grouping of terms,

$$\mathcal{G}\frac{\partial U}{\partial \Psi_k} + R' = 0, \tag{2.30}$$

where

$$\mathcal{G} = \left(\frac{\partial \Psi_k}{\partial t'} + \sum_{i=1}^{n} \frac{\partial \varphi_i}{\partial t}\frac{\partial \Psi_k}{\partial \varphi_i}\right) A_o(U,\underline{x},t) + \sum_{i,j=1}^{n}\left(\frac{\partial \varphi_j}{\partial x_i}\frac{\partial \Psi_k}{\partial \varphi_j} A_i(U,\underline{x},t)\right) \tag{2.31}$$

and R' is the correspondingly transformed vector R from equation (2.6). Now from the properties of Jacobian matrices the coefficient matrix $\mathcal{G}$ may be rewritten as

$$\mathcal{G} = \left(\frac{\partial \Psi_k}{\partial t} A_o(U,\underline{x},t) + \sum_{i=1}^{n} \frac{\partial \Psi_k}{\partial x_i} A_i(U,\underline{x},t)\right). \tag{2.32}$$

Thus the condition that $\partial U/\partial \Psi_k$ should be determined by (2.30) is that $\det \mathcal{G} \neq 0$, while the characteristic manifolds themselves will be determined by the condition det $\mathcal{G} = 0$. The exterior derivatives $\partial U/\partial\Psi_k$ and $\partial\varphi_k$ at any point P on a non-characteristic manifold $\mathcal{S}$ are, of course, equal apart perhaps for their sign. Hence the matrix $\mathcal{G}$ leads to precisely the same characteristic polynomial $Q(P;\underline{\nu};\lambda)$ in (2.12) as does matrix Λ in (2.7), and thus to the same m characteristic manifolds associated with the respective m zeros of $Q(P;\underline{\nu};\lambda)$. The invariance of the characteristic manifolds with respect to a change of coordinates has thus been established. Although for convenience the time variable has not been changed in this particular form of proof this in no way invalidates the proof of invariance. It could for this purpose equally well have been changed arbitrarily along with the other variables involved.

2.4 Characteristic Manifolds As Transporters of Discontinuities of Derivatives

It has already been established in Section 2.1 that the condition the normal derivative $\partial U/\partial\varphi_k$ to the manifold $\mathcal{S}$ with equation $\varphi_k(\underline{x},t) = \text{const.}$ in $\mathbb{R}^{n-1} \times t$ should be obtainable from (2.1) is that the matrix Λ in (2.7) be non-singular. The characteristic manifolds $\mathcal{C}$ for system (2.1) are then those for which this condition does not hold, so that for them

$$\det \Lambda = Q(P;\underline{\nu},\lambda) = 0. \tag{2.33}$$

On a characteristic manifold, as remarked on page 47, the normal derivative $\partial U/\partial\varphi_k$ may be different on the opposite sides $\mathcal{C}_+$ and $\mathcal{C}_-$ of $\mathcal{C}$. The only constraint on this derivative is the one already noted on page 47 which is implied by (2.14). This is that $\partial U/\partial\varphi_k$ may be assigned arbitrarily on adjacent sides of $\mathcal{C}$ provided only that on either side it is proportional to the right eigenvector r in (2.14) that corresponds to the characteristic manifold that is involved. The solution vector U itself must, however, be continuous across this characteristic manifold as it is only the exterior derivative in (2.6) that becomes indeterminate on $\mathcal{C}$. Hence the right eigenvector r associated with $\mathcal{C}$ will be continuous across the characteristic manifold because its elements will depend only on U, $\underline{x}$ and t, which are continuous.

Now when systems (2.1) are considered which are hyperbolic in the t-directio it follows from the implicit function theorem and the definition of λ in (2.9) that provided $\lambda \neq 0$, the equation $\varphi_k(\underline{x},t) = \text{const.}$ for $\mathcal{C}$ may be solved for t and written in the form

$$\Phi_k(\underline{x}) - t = 0. \tag{2.34}$$

So far any given $t = t_1$(const.) a spatial characteristic sub-manifold $\mathscr{T}(t_1)$ exists in $\mathbb{R}^{n-1}$ across which the normal spatial derivative of U may be assigned differently on the opposite sides $\mathscr{T}_+(t_1)$ and $\mathscr{T}_-(t_1)$. This spatial sub-manifold will occupy different positions in $\mathbb{R}^{n-1}$ at different times t so that it may be regarded as a propagating wave $\mathscr{T}(t)$. If the directional derivative $\partial U/\partial \nu$ along the spatial unit normal $\underline{\nu}$ to $\mathscr{T}(t)$ is initially discontinuous across $\mathscr{T}(t)$, then as the only sub-manifolds across which this is possible are characteristic, it must follow that this discontinuity in normal derivative $\partial U/\partial \nu$ will propagate with $\mathscr{T}(t)$. In this case, although a solution manifold will be continuous across $\mathscr{T}(t)$, it will exhibit a discontinuous gradient across it. This situation was illustrated in the simple case of the space $\mathbb{R}^1 \times t$ in Fig.1.

The sides $\mathscr{T}_{\pm}(t)$ of this sub-manifold are determined most naturally by associating with the $\pm$ signs those sides of $\mathscr{T}(t)$ on which the respective waves corresponding to $\mathscr{T}(t \pm \epsilon)$ lie for $\epsilon > 0$. As there are m real zeros $\lambda^{(1)}, \lambda^{(2)}, \ldots, \lambda^{(m)}$ of the characteristic polynomial $Q(\underline{P}; \underline{\nu}, \lambda)$, each associated with a spatial characteristic sub-manifold, it follows that a strictly hyperbolic system (2.1) will have m distinct spatial characteristic sub-manifolds. When the system (2.1) is merely hyperbolic these will not all be distinct.

It is not necessary that $\partial U/\partial \nu$ should be discontinuous across a spatial characteristic sub-manifold, because it is also consistent with (2.14) that on each side of the sub-manifold the derivative $\partial U/\partial \nu$ should be equal to the same scalar multiple of the appropriate right eigenvector r. When this occurs the solution manifold will be smooth across the spatial characteristic sub-manifold.

We have thus demonstrated that for hyperbolic systems (2.1) the characteristic manifolds act as transporters of discontinuities of first order derivatives of the solution vector U. This is an important result because it implies that not only continuous initial data, but also initial data with bounded discontinuities in its derivatives along lines on the initial manifold $\mathscr{S}$ will be transported along spatial characteristic sub-manifolds. A rigorous proof of this result for a general system (2.1) has yet to be given, but in the case of $\mathbb{R}^1 \times t$ there have been a number of fundamental contributions. We mention here only the basic papers by Friedrichs [36], Courant and Lax [37], Douglis [38], Hartman and Wintner [39] and Lax [40]. One of the essential results established in these, which is stronger than the one we have obtained above is that except at a set of points on the initial line having measure zero the characteristic curves through points on the initial line act as transporters of both continuous and Lipschitz continuous data for as long as the solution remains Lipschitz continuous. The proof, which is lengthy, is complicated by the fact that the behaviour of the solution needs to be discussed along the characteristic curves that exist in $\mathbb{R}^1 \times t$, and for a quasilinear hyperbolic system these also depend on the solution.

Propagation of a discontinuity along a characteristic manifold can thus be expected to change its nature when the solution vector ceases to be Lipschitz continuous. A subsequent chapter will be devoted to a detailed study of this question in $\mathbb{R}^1 \times t$ and to the determination of where and when along a characteristic curve transporting a discontinuity in a derivative the solution vector first ceases to be Lipschitz continuous.

Returning now to the case of system (2.1) in $\mathbb{R}^n \times t$ it is important to recognise that as there are m zeros $\lambda^{(i)}$ of the characteristic polynomial $Q(P;\underline{\nu},\lambda)$ there will be m essentially different types of spatial characteristic sub-manifold or wave associated with a hyperbolic system of this type. The different properties possessed by each of these m sub-manifolds follows from the fact that if system (2.1) is strictly hyperbolic, or even only hyperbolic, it will still have associated with it m different eigenvectors $r^{(i)}$ each characterising a different type of wave.

At this stage a geometrical interpretation of the zeros $\lambda^{(i)}$ of $Q(P;\underline{\nu};\lambda)$ will be useful. First, let us recall that as (2.12) was arrived at by seeking the derivative $\partial U/\partial\varphi_k$ normal to $\varphi_k(\underline{x},t) = \text{const.}$, it follows that the vector $\underline{\nu}$ in $Q(P;\underline{\nu};\lambda^{(i)}) = 0$ is to be interpreted as a normal to the spatial characteristic sub-manifold $\mathscr{T}^{(i)}$ associated with $\lambda^{(i)}$. Hence by identifying each $\varphi_i(\underline{x},t) = \text{const.}$ for $i = 1,2,\ldots,n$ in the new coordinate system $\underline{\varphi} = (\varphi_1,\varphi_2,\ldots,\varphi_n)$ with a spatial characteristic sub-manifold $\mathscr{T}^{(i)}$ we obtain a semi-characteristic coordinate system in $\mathbb{R}^n \times t$ in which all variables but the time variable in (2.1) have been transformed.

Let us now denote by $\varphi(\underline{x},t) = \text{const.}$ the equation of any one of these spatial characteristic sub-manifolds $\mathscr{T}$. Then taking the total derivative of $\varphi(\underline{x},t)$ in this sub-manifold gives the result

$$\frac{d\varphi}{dt} = \frac{\partial\varphi}{\partial t} + \frac{\partial\varphi}{\partial x_1}\frac{dx_1}{dt} + \frac{\partial\varphi}{\partial x_2}\frac{dx_2}{dt} + \ldots + \frac{\partial\varphi}{\partial x_n}\frac{dx_n}{dt} = 0 \qquad (2.35)$$

or, more simply still,

$$\frac{-\partial\varphi/\partial t}{|\nabla_x \varphi|} = \underset{\sim}{\nu} \cdot \frac{d\underset{\sim}{x}}{dt} \tag{2.36}$$

where, as before $\underset{\sim}{\nu} = (\nabla_x \varphi)/|\nabla_x \varphi|$ is the unit normal to $\mathcal{T}$ and

$$|\nabla_x \varphi| = \left\{ \sum_{i=1}^{n} \left(\frac{\partial \varphi}{\partial x_i} \right)^2 \right\}^{\frac{1}{2}} . \tag{2.37}$$

From the definition of λ in (2.9) it then follows that (2.36) may be written

$$\lambda = \underset{\sim}{\nu} \cdot \frac{d\underset{\sim}{x}}{dt} . \tag{2.38}$$

As $d\underset{\sim}{x}/dt$ may be interpreted as the velocity of propagation of the spatial characteristic sub-manifold $\mathcal{T}$ at $(\underset{\sim}{x}, t)$, and $\underset{\sim}{\nu}$ is a unit normal to $\mathcal{T}$, the quantity λ is to be interpreted as the velocity of propagation of $\mathcal{T}$ along the normal $\underset{\sim}{\nu}$ to $\mathcal{T}$. Each of the zeros $\lambda^{(i)}$ of $Q(P;\underset{\sim}{\nu};\lambda)$ is then seen to be the speed of propagation of the associated spatial characteristic sub-manifold along a normal $\underset{\sim}{\nu}$ at P. An account of waves and wavefronts employing these ideas is to be found in the paper by Duff [41] who considered the linear case.

If desired a local tangent hyperplane Π to $\mathcal{T}$ may be constructed at any point P. This follows by observing that as $\underset{\sim}{\nu}$ is a normal to $\mathcal{T}$, the hyperplane Π normal to $\underset{\sim}{\nu}$ at P must be tangent to $\mathcal{T}$. The normal vectors $\underset{\sim}{\nu}$ at P define characteristic directions and form the generators of a cone with its vertex at P while the planes Π envelop $\mathcal{T}$.

A similar argument establishes that higher order derivatives $\partial^r U/\partial\varphi^r$ normal to a spatial characteristic sub-manifold may also be discontinuous across it and will propagate with it. This is an idea we shall return to later in connection with systems of equations in $\mathbb{R}^1 \times t$.

The ideas in the present chapter derive from the Cauchy problem for a quasilinear hyperbolic system (2.1) for which to determine the characteristic manifolds it is necessary to know the solution. When linear equations are involved a considerable simplification results in these arguments. Not only may the characteristic manifolds then be determined independently of the solution, but linear superposition may be used to construct the solution itself. A special case is when constant coefficient equations are involved and for an account of the form of solution it is frequently possible to obtain by appeal to the use of plane waves and spherical means we refer to the work by John [42]. Also of relevance to this last problem is the work of John in connection with non-admissible data [43]. Further information is also to be found in Hörmander [1] and Mikhlin [4].

2.5 Characteristic Fields of First Order Quasilinear Hyperbolic Systems in One Space Dimension and Time

Let us now consider the important case of first order quasilinear hyperbolic systems (2.1) in $\mathbb{R}^1 \times t$. The geometrical complexity of characteristic manifolds in many dimensions then reduces to the study of two-dimensional characteristic curves in the (x,t)-plane.

The most general first order quasilinear system of type (2.1) in the (x,t)-plane is

$$A_o(U,x,t)\frac{\partial U}{\partial t} + A_1(U,x,t)\frac{\partial U}{\partial x} + F(U,x,t) = 0, \tag{2.39}$$

where U is a column vector with the n components $u_1, u_2, \ldots, u_n$, A_o and A_1 are $n \times n$ matrices that may depend on x, t and U and F is also an n component column vector with elements that may depend on x, t and U.

Then if (2.39) is hyperbolic in the t-direction it follows that A_o is positive definite, and so has an inverse A_o^{-1}. Pre-multiplying (2.39) by A_o^{-1} then reduces the equation to the more convenient form with which we shall normally work,

$$\frac{\partial U}{\partial t} + A(U,x,t)\frac{\partial U}{\partial x} + B(U,x,t) = 0, \tag{2.40}$$

where $A = A_o^{-1}A_1$ and $B = A_o^{-1}F$.

As there is now only one space dimension involved the unit normal vector $\underline{\nu}$ involved in the definition of the characteristic polynomial simplifies to the unit vector in the x-direction so that $\nu_1 = 1$ in (2.12). The characteristic polynomial for (2.40) now no longer depends on $\underline{\nu}$ and so will be denoted by $Q(P;\lambda)$ where

$$Q(P;\lambda) \equiv |A - \lambda I| = 0, \tag{2.41}$$

and since (2.40) is assumed to be hyperbolic the n eigenvalues $\lambda^{(1)}, \lambda^{(2)}, \ldots, \lambda^{(n)}$ of A will be real and the n right eigenvectors $r^{(1)}, r^{(2)}, \ldots, r^{(n)}$

will span E^n, where

$$Ar^{(i)} = \lambda^{(i)} r^{(i)} \quad \text{for } i = 1,2,\ldots,n. \tag{2.42}$$

We notice that when defining hyperbolicity in Definition 2.1, the condition that the right eigenvectors $r^{(i)}$ should span the space to which Λ belongs could equally well be replaced by the condition that the left eigenvectors $\ell^{(i)}$ span the space. These vectors $\ell^{(i)}$ are defined by the relations

$$\ell^{(i)}A = \lambda^{(i)}\ell^{(i)} \quad \text{for } i = 1,2,\ldots,n, \tag{2.43}$$

which correspond to (2.42).

In the (x,t)-plane equation (2.38) simplifies to

$$\frac{dx}{dt} = \lambda(U,x,t), \tag{2.44}$$

so that as there are n eigenvalues $\lambda^{(i)}$ of A there will be n such differential equations (2.44). At this point is is convenient to introduce a parameter σ into (2.44) in a specially simple way by writing

$$\frac{dx}{d\sigma} = \lambda(U,x,t), \quad \frac{dt}{d\sigma} = 1, \tag{2.45}$$

thereby assigning a direction vector $(\lambda,1)$ to points $x = x(\sigma)$, $t = t(\sigma)$ of the (x,t)-plane. As a direction is associated with each point of the plane this leads naturally to the related concepts of a characteristic field and a family of characteristic curves, the first of which we now define.

Definition 2.2 (Characteristic Field)
The direction field $(\lambda^{(i)},1)$ associated with the i-th eigenvalue $\lambda^{(i)}$ of A is defined over the (x,t)-plane at points $x = x(\sigma)$, $t = t(\sigma)$ by the equations

$$\frac{dx}{d\sigma} = \lambda^{(i)}(U,x,t), \quad \frac{dt}{d\sigma} = 1. \tag{2.46}$$

It is called the i-th characteristic field of the hyperbolic system (2.40). ▮

It is apparent from this definition that when system (2.40) is strictly hyperbolic there will be n distinct characteristic fields associated with it. If for some specific $\lambda^{(i)}$ equations (2.46) are integrated with respect to σ, subject to the initial conditions $x(0) = \xi$, $t(0) = \tau$, say, they will define a curve in the (x,t)-plane with direction vector $(\lambda^{(i)},1)$ at the point

$x = x^{(i)}(\sigma,\xi,\tau)$, $t = \tau + \sigma$. This curve is called the i-th characteristic curve $C^{(i)}_{(\xi,\tau)}$ of the hyperbolic system (2.40) through the point (ξ,τ) in the (x,t)-plane. In the quasilinear case since $\lambda^{(i)}$ is, in general, a function of U, the determination of characteristics requires knowledge of the solution. The i-th characteristic field introduced via the parameterisation σ then associates a direction along $C^{(i)}_{(\xi,\tau)}$ which is tangent to it at each point and which corresponds to the sense of increasing σ.

Definition 2.3 (Characteristic Curves)
The family of curves $C^{(i)}$ that results from the solution of the pair of differential equations

$$\frac{dx}{d\sigma} = \lambda^{(i)}, \quad \frac{dt}{d\sigma} = 1 \tag{2.47}$$

is called the i-th family of characteristic curves associated with the hyperbolic system (2.40). ▮

As with characteristic fields, there will only be n distinct families of characteristic curves associated with system (2.40) when it is strictly hyperbolic. These families of characteristic curves are of considerable importance when studying system (2.40) because they feature both in its solution and in the way problems must be posed for such systems. Before examining these ideas further let us first see how they serve to simplify system (2.40) itself.

Two equivalent methods of approach will now be outlined for the simplification of system (2.40), both of which employ the concept of characteristic curves. The first of these is most useful for theoretical considerations of hyperbolic systems, while the second is better suited to the practical problems that arise when studying solutions to particular systems. We begin the theoretical reduction of system (2.40) to what is called standard characteristic form by recalling that as the system is assumed to be hyperbolic, the left eigenvectors $\ell^{(i)}$ of A span E^n.

On account of this fact the matrix A may be written

$$A = L D L^{-1}, \tag{2.48a}$$

where the non-singular $n \times n$ matrix L has columns comprising the right eigenvectors $r^{(1)}, r^{(2)}, \ldots, r^{(n)}$, so that

$$L = [\, r^{(1)} \,\vdots\, r^{(2)} \,\vdots\, \ldots \,\vdots\, r^{(n)} \,] \tag{2.48b}$$

while D is the n X n diagonal matrix

$$D = \begin{bmatrix} \lambda^{(1)} & 0 & \cdots & 0 & 0 \\ 0 & \lambda^{(2)} & \cdots & 0 & 0 \\ \cdot & \cdot & \cdots & 0 & \cdot \\ \cdot & \cdot & \cdots & \lambda^{(n-1)} & \cdot \\ 0 & 0 & \cdots & 0 & \lambda^{(n)} \end{bmatrix}. \tag{2.48c}$$

The change to the n element column vector V, where U = LV, then brings about the desired simplification, for system (2.40) becomes

$$V_t + DV_x + (L^{-1}L_t + DL^{-1}L_x)V + L^{-1}B = 0. \tag{2.49}$$

This is now a quasilinear equation for V which is hyperbolic in the t-direction because the eigenvalues of D are merely $\lambda^{(1)}, \lambda^{(2)}, \ldots, \lambda^{(n)}$ and the eigenvectors span E^n. The simplification that has been achieved comes about because each element of vector V in the first two terms of (2.49) is now differentiated in a different direction; in fact at any point (x_o, t_o) the i-th element v_i is differentiated in the direction $(\lambda^{(i)}, 1)$ along the characteristic curve $C^{(i)}$ passing through (x_o, t_o). To see this it is only necessary to display the i-th equation of (2.49) which is

$$\frac{\partial v_i}{\partial t} + \lambda^{(i)}(V, x, t)\frac{\partial v_i}{\partial x} + g_i = 0, \tag{2.50}$$

where g_i is the i-th element of $(L^{-1}L_t + DL^{-1}L_x)V + L^{-1}B$. It then follows from what has just been established, or alternatively from the work of Section 1.6, that v_i is differentiated along a member of the i-th family of characteristics $C^{(i)}$ in the direction $(\lambda^{(i)}, 1)$.

An alternative simplification of system (2.40) comes about when it is premultiplied by $\ell^{(i)}$ and the result $\ell^{(i)}A = \lambda^{(i)}\ell^{(i)}$ from (2.43) is used, for it then reduces to

$$\ell^{(i)}(U_t + \lambda^{(i)}U_x) + b_i = 0 \tag{2.51}$$

with $b_i = \ell^{(i)}B$. This is a scalar equation; but in the expression $(U_t + \lambda^{(i)}U_x)$ it is now the vector U that is differentiated along a member of the i-th family of characteristics $C^{(i)}$ in the direction $(\lambda^{(i)},1)$. Expressed another way, equation (2.51) defines the linear combination of the scalar equations in (2.40) that must be formed if in the resulting equation each element of vector U is to be differentiated in the same direction. The linear combination defined by the i-th left eigenvector then corresponds to diferentation of U along a characteristic belonging to the i-th family of characteristics. Result (2.51) which is true for $i = 1,2,\ldots,n$ along curves belonging to the respective families of characteristics $C^{(i)}$ is less simple in form than (2.50) but more convenient to use in practice.

In the (x,t)-plane Cauchy data will now be specified on an arc Γ and the concept of a characteristic field then makes the classification of this arc as either space-like or time-like particularly easy. Through any point P in the (x,t)-plane there will be n characteristic curves $C_P^{(1)}, C_P^{(2)},\ldots,C_P^{(n)}$, which will all be distinct in the strictly hyperbolic case. To each curve $C_P^{(i)}$ we now associate the direction corresponding to the direction vector $(\lambda_P^{(i)},1)$. Then if we draw from P only that part of each curve $C_P^{(i)}$ which corresponds to σ increasing we arrive at a fan of characteristic curves radiating out from P. If the eigenvalues at P are now ordered so that

$$\lambda_P^{(n)} < \lambda_P^{(n-1)} < \ldots < \lambda_P^{(2)} < \lambda_P^{(1)} , \qquad (2.52)$$

then the extreme left hand characteristic arc will correspond to $\lambda_P^{(n)}$ and the

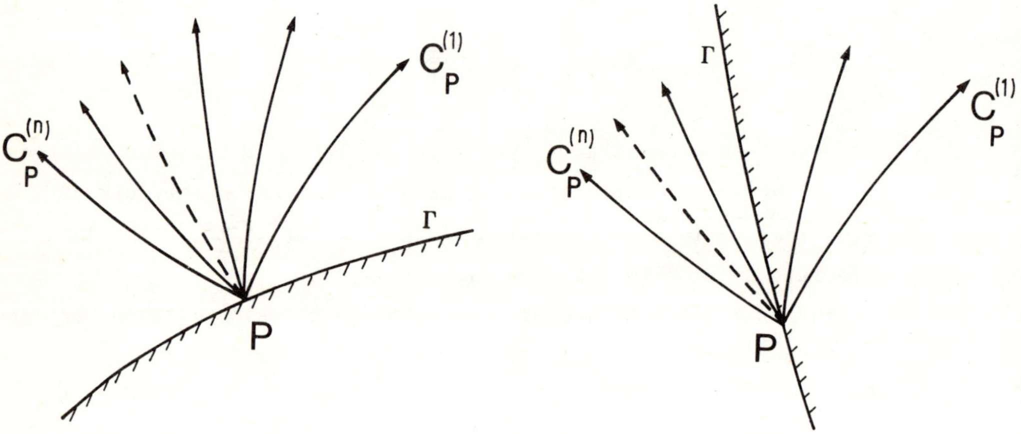

(a) Time-like arc Γ at P

(b) Space-like arc Γ at P

Classification of arcs Γ in the (x,t)-plane

Fig. 7.

extreme right hand one to $\lambda_P^{(1)}$. Here multiple eigenvalues are understood to be numbered according to their multiplicity.

The tangents to these two characteristic arcs then define the two extreme directions through P between which an arc Γ may be drawn and still be space-like, as are the tangents drawn at P to the characteristics $C_P^{(n-1)}, C_P^{(n-2)}, \ldots, C_P^{(2)}$. Any arc Γ through P that does not lie between these two directions will be time-like. These situations are illustrated in Fig. 7.

The ordering of the eigenvalues in (2.52) enables the terms time-like and space-like to be defined simply with respect to an arc Γ. However as in quasilinear hyperbolic systems the eigenvalues $\lambda^{(i)}$ depend on U, as well as on x and t, this classification may only be made locally in the neighbourhood of an arc Γ on which the solution is known. In the linear case no such problem arises since the characteristic curves may be determined from (2.47) without reference to the solution. We thus arrive at the following definition which is a pointwise definition for quasilinear systems (2.40) and global for linear ones.

<u>Definition 2.4</u> (Space-like and Time-like Arcs)
If an arc Γ passing through point P of the (x,t)-plane has the direction vector $(\mu,1)$ at that point, then it will be space-like at P with respect to system (2.40) if

$$\lambda_P^{(n)} \leq \mu \leq \lambda_P^{(1)} \quad ;$$

otherwise it will be time-like at P. ▌

The situation may arise that some segments of an arc Γ are time-like while others are space-like. This happens with mixed boundary and initial value problems which will be mentioned in the next section. To appreciate the way data must then be assigned to Γ in order that a unique solution should exist it is usual to work with the standard characteristic form (2.49) to which (2.40) may always be reduced. This, then, will be the starting point for what follows.

2.6 Initial Value Problems and Mixed Initial and Boundary Value Problems

Let us consider an initial value problem for a strictly hyperbolic system (2.40) when it has been reduced to the standard characteristic form

$$V_t + D V_x + G = 0 \tag{2.53}$$

given in (2.49), where D is the diagonal matrix of eigenvalues (2.48c) and $G = (L^{-1}L_t + DL^{-1}L_x)V + L^{-1}B$. Our task will now be to discuss under what conditions a unique solution V exists at a point P in the (x,t)-plane in terms of Cauchy data specified on a time-like arc. It will suffice for us to accomplish this task when the time-like arc is, in fact, simply the x-axis itself. The Cauchy data on t = 0 will then take the form

$$V(x,0) = V_o(x). \qquad (2.54)$$

Let us suppose a solution V exists at point P. Then through P will pass a fan of n characteristic curves $C^{(1)}, C^{(2)}, \ldots, C^{(n)}$ associated with $\lambda^{(1)}, \lambda^{(2)}, \ldots, \lambda^{(n)}$, respectively. If these are traced backwards they will intersect the initial line t = 0 at points $Q_1, Q_2, \ldots, Q_n$ as shown in Fig. 8. For reasons that will soon become apparent the region $\mathscr{D}$ between the two extreme characteristics $C^{(1)}$ and $C^{(n)}$ through P and the line t = 0, shown as the shaded region in Fig. 8, is called the domain of determinacy of the solution with respect to the Cauchy data specified on the interval $Q_1 Q_n$. The interval itself between Q_1 and Q_n on the initial line t = 0 which we will denote by Δ_P is called the domain of dependence of the solution at P.

The domain of dependence Δ_R of any point R in the region $\mathscr{M}$ exterior to $\mathscr{D}$ at which a solution also exists must be such that $\Delta_R \not\subset \Delta_P$. This follows because by assuming a solution to exist at both P and R, the Jacobian of the transformation U = LV must be non-vanishing, at least until P and R are reached, so that the backward drawn characteristics of any one family through P and R cannot intersect. Depending on the location of points P and R, so the sets of points comprising the two domains of dependence Δ_P and Δ_R may, or may not, be disjoint.

We now return to consideration of the solution V at P, and convert the system of differential equations (2.53) to a system of integral equations. To achieve this, we employ the fact that has already been noted, that the i-th equation in (2.53) has v_i differentiated only along a characteristic belonging to the i-th family $C^{(i)}$. So, integrating each such equation along the appropriate characteristic arc $Q_i P$, we obtain the result

$$v_i(P) = v_i(Q_i) + \int_{Q_iP} g_i(V)dt\ , \quad \text{for } i = 1,2,\ldots,n, \qquad (2.55)$$

where g_i is the i-th element of G and $\int_{Q_iP}$ denotes an integral along the characteristic arc Q_iP. We thus arrive at a system of n Volterra integral equations (2.55) in place of the original system (2.53). It should be remarked here that a modified system results when (2.40) is merely hyperbolic, since then there are no longer n distinct characteristic arcs Q_iP, though the elements $g_i(V)$ will be different because the eigenvectors $r^{(i)}$ of D span E^n.

The determination of the solution to this problem is easiest in the case when system (2.40) has the weak form of nonlinearity that is classified as being semi-linear. That is to say when the matrix A in (2.40) is independent of U but the vector B depends nonlinearly on the solution vector U. In terms of system (2.53) this is equivalent to matrix D being independent of V but G depending nonlinearly on V. Here the matrices A and D may still depend on x and t.

This problem then has a solution if it can be shown by iteration that the transformation $\tilde{V} = \mathcal{L}(V)$, where

$$\tilde{v}_i(P) = v_i(Q_i) + \int_{Q_iP} g_i(V)dt, \text{ for } i = 1,2,\ldots,n, \tag{2.56}$$

has a fixed point, with $\tilde{v}_i$ the i-th element of $\tilde{V}$. We omit the details of this proof which amounts to showing that the transformation $\mathcal{L}$ is a contraction mapping. However, once this has been achieved, the fixed point property of $\mathcal{L}$ then establishes both the existence and the uniqueness of the solution V at P. The result so established may then be used to extend the existence and uniqueness proof to the quasilinear case by a further process of iteration. This is accomplished by replacing system (2.53) by

$$V_t^{r+1} + D(V^r,x,t)V_x^{r+1} + G(V^r,x,t) = 0, \tag{2.57}$$

where V^r denotes the r-th iterate for the solution vector V and

$$V^{r+1}(x,0) = V_o(x). \tag{2.58}$$

For the details of the arguments involved we refer to the papers by Friedrichs [36] and Lax [40].

The reasons for the names domains of dependence and determinacy are now apparent. At point P the unique solution V depends only on the Cauchy data on the interval Δ_P between points Q_1 and Q_n on the initial line t = 0, which it is thus appropriate to call the domain of dependence. By this same argument, the solution at all points of the domain of determinacy $\mathscr{D}$ will be determined uniquely by the Cauchy data given on Δ_P. It is a direct consequence of these conclusions that the Cauchy data on t = 0 outside the interval Δ_P can have no effect on the solution within $\mathscr{D}$.

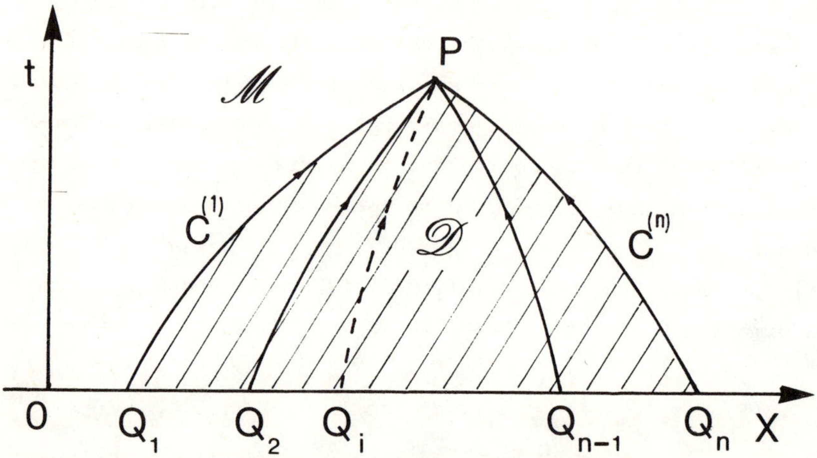

Fig. 8. Domains of dependence of determinacy.

The Cauchy data at any single point Q of the initial line will have no domain of determinacy, though it will influence the solution at points of a region $\mathscr{R}$ that occupies part of the (x, t)-plane. This region is known as the range of influence of Q and it is the region contained between the extreme characteristics $C^{(1)}$ and $C^{(n)}$ radiating out from Q. The region is so named because, as all the characteristics issuing out from Q lie within $\mathscr{R}$, no point in the (x, t)-plane outside $\mathscr{R}$ can be influenced by the Cauchy data at Q. Fig. 9. illustrates the situation and shows $\mathscr{R}$ as a shaded region.

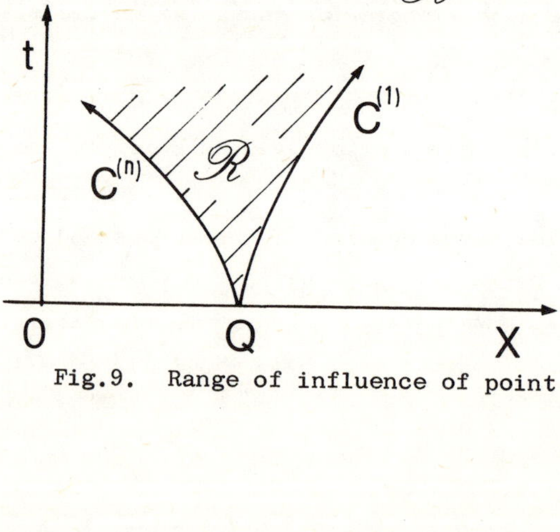

Fig. 9. Range of influence of point Q.

When the details of this proof are followed through, and the result is then interpreted in terms of the original system (2.40), we arrive at the following theorem essentially due to Friedrichs [36] and Lax [40].

Theorem 2.1 (Existence and Uniqueness in the General Quasilinear Case) Let the quasilinear system of partial differential equations

$$U_t + A(U,x,t)U_x + B(U,x,t) = 0 \tag{2.59}$$

be such that:

i) it is hyperbolic in the t-direction throughout all the (x,t)-plane;

ii) the coefficient matrices $A(U,x,t)$, $B(U,x,t)$ have Lipschitz continuous partial derivatives;

iii) it satisfies an initial condition

$$U(x,0) = U_o(x), \quad a < x < b \tag{2.60}$$

for which dU_o/dx is Lipschitz continuous.

Then a unique solution to (2.59) and (2.60) exists in the neighbourhood of the interval $a < x < b$ of the initial line $t = 0$ and, furthermore, within this neighbourhood the solution has Lipschitz continuous partial derivatives. ∎

This theorem only ensures the existence of a unique solution for a finite elapsed time beyond $t = 0$. To determine precisely how the solution behaves if non-uniqueness occurs, and whether or not it can be extended beyond this time in any meaningful way, requires further analysis. These topics are of central importance to these research notes and will form the body of Chapters 4 and 5. That Theorem 2.1 is inadequate to ensure global existence in the case of a single equation, even when the coefficients and data are analytic, has already been made clear in connection with Fig.5 of Section 1.6.

More complicated than the pure initial value problem just discussed is the mixed initial and boundary value problem that results when part of the data is given on a time-like arc and part on a space-like arc. This situation is depicted in Fig.10, in which the arc Σ containing segment Q_1R is supposed to be purely space-like, while the x-axis to the right of point R is time-like. Here the space-like arc Σ bounds the left of the region in which a solution is required. It could also have bounded the right of the region, or arcs Σ_1 and Σ_2 could bound at both left and right, separated by a time-like segment.

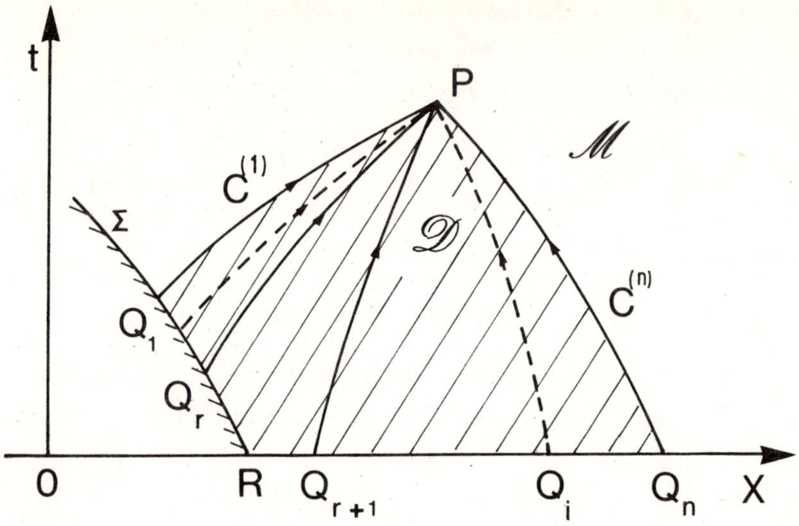

Fig.10. Mixed initial and boundary value problem

The complexity of this problem arises from the fact that the space-like boundary Σ influences the solution in a different way from the time-like boundary. More than this is involved, however, because the data to be specified on Σ cannot be the usual Cauchy data. In order to explore these ideas further we now consider a typical point P in Fig. 10 in the region $\mathscr{M}$ above the x-axis and to the right of a purely space-like arc Σ ,for which r backward drawn characteristics intersect Σ and n-r the initial line. The solution in the region $\mathscr{D}$ then depends on the space-like data between Q_1 and R on Σ and on the time-like data between R and Q_n; in particular on the data at the points $Q_1, Q_2, \ldots, Q_n$.

As in the initial value problem it is again possible to arrive at an equivalent integral equation formulation by integrating system (2.53) along its characteristics. The resulting system of n equations is then

$$v_i(P) = v_i(Q_i) + \int_{Q_iP} g_i(V)dt, \text{ for } i = 1,2,\ldots,n, \qquad (2.61)$$

though it is now necessary to give consideration to the way data is to be specified along Σ .

For this purpose we first take note of the fact that if Σ is purely space-like, then a characteristic may nowhere be tangent to it. Consequently, if P is moved to any point P' on the arc Σ there will still be n-r backward drawn characteristics from P' that intersect the initial line as in Figs. 10 and 11.

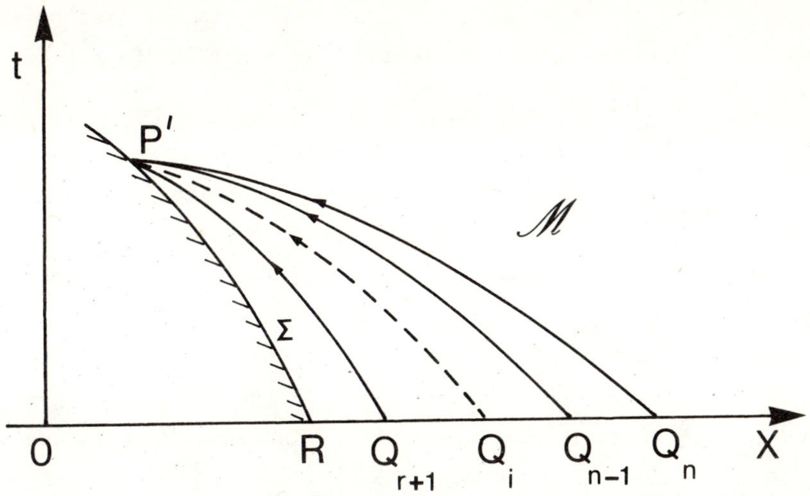

Fig. 11. The n-r backward drawn characteristics from P' on Σ

In the quasilinear case the purely space-like condition on Σ can only be guaranteed a priori for P' close to R. We conclude from this that as initial data on t = 0 can only provide n-r conditions at points P' of Σ, the remaining r conditions on Σ from which to determine the solution both at points P' and at points like P in Fig. 10. must be specified on Σ. Suppose there to be r conditions specified at points P' of Σ of the form

$$
\begin{aligned}
v_1(P') &= f_1(v_{r+1}(P'), \ldots, v_n(P')), \\
v_2(P') &= f_2(v_{r+1}(P'), \ldots, v_n(P')), \\
v_r(P') &= f_r(v_{r+1}(P'), \ldots, v_n(P')).
\end{aligned}
\tag{2.62}
$$

Then by employing results (2.61) along the n-r characteristic arcs in Fig. 11 and iterating it is possible to determine the boundary values $v_{r+1}(P')$, $v_{r+2}(P'),\ldots,v_n(P')$. Once these are known the remaining boundary values $v_1(P')$, $v_2(P'),\ldots,v_r(P')$ may be determined from equations (2.62). When all the boundary values have been determined along Σ, equations (2.61) may then be used to determine the solution V(P) for points P interior to $\mathscr{M}$.

By employing Theorem 2.1 we are thus able to arrive at the following theorem concerning the proper formulation of a mixed initial and boundary value problem. In doing so we take account of the fact that the space-like arc Σ may either lie to the left of the region $\mathcal{M}$ in which a solution is required, as in Fig. 10, or to the right. Indeed there may also be such a space-like arc to both the left and right of $\mathcal{M}$ separated by a segment of time-like arc.

Theorem 2.2 (Mixed Initial and Boundary Value Problem)
Let a space-like arc Σ issuing out from a point R on the initial line comrise part of the boundary of a region $\mathcal{M}$ in which a solution to system (2.53) is required. Let there be n-r characteristics which when traced backwards through $\mathcal{M}$, from any point P' of Σ that is close to R, intersect the initial line on which the Cauchy data $V(x,0) = V_o(x)$ is specified. Then for a unique Lipschitz continuous solution to exist in some neighbourhood of the time-like part of boundary of $\mathcal{M}$ it is sufficient that r independent relations be given at points P' along Σ of the form

$$v_1(P') = f_1(v_{r+1}(P'), \ldots, v_n(P')),$$

$$v_2(P') = f_2(v_{r+1}(P'), \ldots, v_n(P')),$$

$$\cdots\cdots\cdots$$

$$v_r(P') = f_r(v_{r+1}(P'), \ldots, v_n(P')),$$

and that the matrix D, the vectors G and V_o and the functions $f_1, f_2, \ldots, f_r$ have Lipschitz continuous derivatives. ∎

Before concluding this section it is necessary to take note of the fact that the functions $f_1, f_2, \ldots, f_r$ are not completely arbitrary. To show the nature of the constraint on them let us denote differentiation of v_i along Σ by $d\Sigma/d\sigma$, when it is a consequence of equations (2.62) that

$$\frac{dv_i}{d\sigma} = \frac{\partial f_i}{\partial v_{r+1}} \frac{dv_{r+1}}{d\sigma} + \frac{\partial f_i}{\partial v_{r+2}} \frac{dv_{r+2}}{d\sigma} + \ldots + \frac{\partial f_i}{\partial v_n} \frac{dv_n}{d\sigma} \qquad (2.63)$$

for $i = 1, 2, \ldots, r$. Then as the derivatives with respect to σ may be computed from the assigned data at corner points like R, where Σ meets the initial line, it follows that the functions $f_1, f_2, \ldots, f_r$ must be required to satisfy (2.63) at any such points in order that they are compatible with the Cauchy data. We take note here that the space-like arc Σ may move with time.

2.7 Examples of Characteristics, Initial and Boundary Value Problems

This section provides some typical examples of the material that has been discussed so far. For simplicity of exposition the examples have been confined to the case of one space dimension and time.

Example 1.

Characteristic Curves for the Tricomi Equation

The Tricomi equation (2.24a) is a linear second order partial differential equation which may be expressed as the first order system (2.24b). The two families of characteristic curves $C^{(1)}$ and $C^{(2)}$ may be found by solving the equations corresponding to (2.47). Making the variable changes $t \to x$ and $x \to y$ to bring the notation in (2.47) into agreement with that in system (2.24b) we obtain the result

$$C^{(i)} : \quad \frac{dy}{d\sigma} = \lambda^{(i)}, \quad \frac{dx}{d\sigma} = 1 \quad \text{for } i = 1,2, \tag{2.64}$$

where the $\lambda^{(1)}$, $\lambda^{(2)}$ are determined by $|A - \lambda I| = 0$, or from (2.24c) by

$$\begin{vmatrix} -\lambda & x \\ -1 & -\lambda \end{vmatrix} = 0.$$

As already noted on page 53, the $\lambda^{(i)}$ are only real when $x < 0$, which corresponds to the region in which the Tricomi equation is hyperbolic. We shall set $\lambda^{(1)} = \sqrt{-x}$ and $\lambda^{(2)} = -\sqrt{-x}$ for $x < 0$. Then integration of (2.64) gives at once

$$\begin{aligned} C^{(1)} &: \quad y = c - \frac{2}{3}(\sqrt{-x})^3, \\ C^{(2)} &: \quad y = c + \frac{2}{3}(\sqrt{-x})^3, \end{aligned} \tag{2.65}$$

where c is an arbitrary constant of integration which serves as the parameter with which to identify a particular member of the $C^{(1)}$ or $C^{(2)}$ families of cubic parabola characteristic curves. Because of the linearity of the Tricomi equation the characteristic curves are independent of the solution.

Example 2.

Characteristic Curves for the Shallow Water Approximation

The shallow water approximation introduced in system (1.17) was shown on page 53 to have as solution to $|A - \lambda I| = 0$ the results $\lambda = u \pm c$. If we set $\lambda^{(1)} = u + c$ and $\lambda^{(2)} = u - c$, then the two families of characteristic curves $C^{(1)}$ and $C^{(2)}$ (which it is often convenient to denote by $C^{(\pm)}$ corresponding to $\lambda^{(\pm)} = u \pm c$) are given by integrating

$$C^{(1)} : \quad \frac{dx}{dt} = u + c,$$
$$C^{(2)} : \quad \frac{dx}{dt} = u - c. \tag{2.66}$$

This integration is only possible when the solution is known.

Example 3.

Gas Motion in a Closed Tube

Riemann posed a mixed initial and boundary value problem [44] for gas in a tube of finite length. His problem was to determine the one-dimensional motion of a polytropic gas enclosed between two fixed walls at $x = 0$ and $x = \ell$ when the initial density and velocity variation are specified.

The system of equations involved is the one-dimensional form of equations (1.13), and it may be written

$$U_t + AU_x = 0, \tag{2.67a}$$

where

$$U = \begin{bmatrix} \rho \\ u \end{bmatrix} , \quad A = \begin{bmatrix} u & \rho \\ a^2/\rho & u \end{bmatrix}. \tag{2.67b}$$

The two families of characteristic curves $C^{(1)}$, $C^{(2)}$ determined by (2.47) and $|A - \lambda I| = 0$ are given by integrating

$$C^{(1)} : \quad \frac{dx}{dt} = u + a,$$
$$C^{(2)} : \quad \frac{dx}{dt} = u - a. \tag{2.68}$$

The initial conditions to be imposed are arbitrary, subject only to the condition that the gas is stationary at the walls $x = 0$ and $x = \ell$ and that the density is non-negative. Hence we arrive at the

initial conditions

$$u(x,0) = u_o(x) \text{ with } u_o(0) = u_o(\ell) = 0,$$

and

$$\rho(x,0) = \rho_o(x) \text{ with } \rho_o(x) > 0. \tag{2.69a}$$

The boundary conditions to be imposed are simply that gas in contact with the walls remains at rest for all time, so that we arrive at the

boundary conditions

$$u(0,t) \;=\; u(\ell,t) = 0. \tag{2.69b}$$

This problem provides an example of a region bounded on both sides by fixed space-like arcs.

The way the gas motion evolves from the initial conditions and where and when a gas shock forms was first resolved by Ludford [45] and, subsequently, by Jeffrey [46] using a different treatment.

Example 4.

The Piston Problem of Gas Dynamics

A different mixed initial and boundary value problem arises from the so-called piston problem of gas dynamics [9]. It involves the determination of the one-dimensional flow of a polytropic gas in a semi-infinite tube, one end of which is sealed by a piston. There is thus only one boundary condition of a space-like type at the piston wall. In the simplest problem of this type the gas, which is assumed initially to be in equilibrium and at rest, is set in motion by moving the piston.

The system of equations is again the one given in (2.67) but now the initial and boundary conditions are different. If the origin of the x-axis is taken at the piston wall with the positive x-axis being directed into the gas, the initial conditions in the gas will be (the origin is fixed in space)

$$\rho(x,0) = \rho_0 \text{ (const) and } u(x,0) = 0 \quad \text{for } x \geq 0. \tag{2.70a}$$

As the gas in contact with the piston wall will be assumed to move with it, the boundary condition must be that the gas in contact with the piston at time t has the piston velocity at that time. Suppose the piston path, as a function of time, is given by $x = \sigma(t)$. Then the piston speed at time t will be $d\sigma/dt$. The boundary condition on the moving piston wall then becomes

$$u(\sigma(t),t) \;=\; \frac{d\sigma}{dt}\,. \tag{2.70b}$$

As the piston starts from rest we must require of $\sigma(t)$ that $d\sigma/dt = 0$ when $t = 0$. Since initially $u = 0$ the characteristic curves through the origin determined by equations (2.68) will have slopes $\pm a_o$, where a_o is the speed of sound in the equilibrium state determined by the initial conditions and the polytropic gas law involved.

The fact that the piston wall is, at least initially, space-like then follows from the fact that the tangent to the piston path at time $t = 0$ (it is vertical when drawn in the (x,t)-plane) lies between the two characteristic curves drawn through the origin. This, then, is an example of a mixed initial and boundary value problem in which the space-like boundary (the piston) is moving.

2.8 Waves Adjacent to a Constant Solution - Characteristic Equations

In Section 2.1 it was established that the jump in the directional derivative of U normal to a characteristic manifold must be proportional to the right eigenvector r corresponding to that manifold. This result takes on a particularly simple and useful form when the characteristic manifold involved bounds on one side a non-constant solution and on the other side a constant one. The purpose of the present section is to make clear for later use the precise form this relation takes.

Suppose the solution on side $\mathcal{S}_-$ is non-constant and the solution on side $\mathcal{S}_+$ is constant, while $\mathcal{S}$ has the equation $\psi = 0$ with ψ parameterised so that $\psi > 0$ on side (+). Then on side (-) close to $\mathcal{S}$ we may write the differential relation

$$dU = \left(\frac{\partial U}{\partial \psi}\right)_{\psi=0-} d\psi. \tag{2.71}$$

Now although U is continuous across $\mathcal{S}$ the directional derivative $\partial U/\partial\psi$ normal to $\mathcal{S}$ is discontinuous and, since $U \equiv U_o$ (const.) on side (+), it follows that

$$\left(\frac{\partial U}{\partial \psi}\right)_{\psi=0+} = 0.$$

Consequently the jump in $\partial U/\partial\psi$ across $\psi = 0$ which we shall denote by Π and define by the relation

$$\Pi \equiv \left(\frac{\partial U}{\partial \psi}\right)_{\psi=0-} - \left(\frac{\partial U}{\partial \psi}\right)_{\psi=0+}, \tag{2.72}$$

and which will be a function of position on the characteristic surface $\mathcal{S}$, reduces to

$$\Pi = \left(\frac{\partial U}{\partial \psi}\right)_{\psi=0-}. \qquad (2.73)$$

A comparison of (2.71) and (2.73) now establishes the important fact that Π is proportional to dU, so we may write

$$\Pi \propto dU. \qquad (2.74)$$

As Π is proportional to the right eigenvector r that is associated with $\mathscr{A}$, the implication of this last result is that if in (2.14) the characteristic manifold with normal at P defined by $(-\lambda^{(j)}, \underline{\nu})$ separates a constant solution from a non-constant one, then across this characteristic manifold we have the result

$$\sum_{i=1}^{n}\{\nu_i A_i(P) - \lambda^{(j)} A_o(P)\}dU = 0. \qquad (2.75)$$

These homogeneous equations are sometimes called the characteristic equations of system (2.1) at point P. They determine the relationships that exist between the differentials of the elements of U of system (2.1) at point P across the j-th type of wave represented by this characteristic manifold. The equations may be regarded as the compatibility relations to be satisfied by the differentials across the j-th wave at point P.

Inspection of the forms of systems (2.1) and (2.75) gives a simple rule by which the characteristic equations for (2.1) may be derived directly from the coefficient matrices of $\partial U/\partial t$ and $\partial U/\partial x_i$. It merely amounts to discarding the vector B in system (2.1) and making the replacements

$$\frac{\partial(\cdot)}{\partial t} \to -\lambda d(\cdot) \quad \text{and} \quad \frac{\partial}{\partial x_i} \to \nu_i d(\cdot).$$

In terms of the spatial gradient operator ∇ this is also equivalent to the replacements

$$\frac{\partial(\cdot)}{\partial t} \to -\lambda d(\cdot) \quad \text{and} \quad \nabla(\cdot) \to \underline{\nu} d(\cdot). \qquad (2.76)$$

To see how these replacements work let us consider the following vector form of the equations of unsteady isentropic compressible gas flow given in equations (1.13):

$$\frac{\partial \rho}{\partial t} + \nabla.(\rho \underset{\sim}{q}) = 0, \qquad (2.77a)$$

$$\frac{\partial \underset{\sim}{q}}{\partial t} + \underset{\sim}{q}.\nabla \underset{\sim}{q} + \frac{a^2}{\rho}\nabla \rho = 0. \qquad (2.77b)$$

Here the relationships $p = p(\rho)$ and $a^2 = \partial p/\partial \rho$ have been used to replace ∇p by $a^2 \nabla \rho$.

Making the replacements indicated in (2.76) then gives for the characteristic equations

$$-\lambda d\rho + \underset{\sim}{\nu}.d(\rho \underset{\sim}{q}) = 0, \qquad (2.78a)$$

$$-\lambda d\underset{\sim}{q} + (\underset{\sim}{q}.\underset{\sim}{\nu})d\underset{\sim}{q} + \frac{a^2}{\rho}\underset{\sim}{\nu} d\rho = 0. \qquad (2.78b)$$

If these are expressed in matrix form it is at once apparent that the coefficient matrix for the vector differential dU with elements $d\rho$, du, dv, dw is the same as the one from which the characteristic polynomial $Q(P;\underset{\sim}{\nu},\lambda)$ on page 49 was derived. In equations (2.77a,b) there was no term corresponding to B in system (2.1) that needed to be discarded when applying the rule.

When system (2.40) is considered, which involves only one space dimension and time, result (2.75) undergoes considerable simplification. On account of its importance this result merits being given formal expression.

Theorem 2.3 (Characteristic Equations in One Space Dimension and Time). Let the system

$$U_t + A(U,x,t)U_x + B(U,x,t) = 0$$

be hyperbolic in the t-direction, and denote by $\lambda^{(1)},\lambda^{(2)},\ldots,\lambda^{(n)}$ the eigenvalues of the $n \times n$ coefficient matrix A. In addition, let a characteristic curve $C^{(i)}$ corresponding to $\lambda = \lambda^{(i)}$ separate a region in which U = const. from one in which U is non-constant. Then the differentials $du_1,du_2,\ldots,du_n$ of the elements $u_1,u_2,\ldots,u_n$ of U satisfy the n homogeneous equations

$$\{A - \lambda^{(i)}I\}dU = 0 \qquad (2.79)$$

across the characteristic curve $C^{(i)}$. ▮

3 Riemann invariants and simple waves

3.1 Riemann Invariants

The notion of a Riemann invariant first arose in connection with gas dynamics [7,9,20] and is associated with a class of particularly simple homogeneous systems of type (2.40) in one space dimension and time. It will be the purpose of this section to examine this class of systems which involve only two dependent variables and in which x and t are not present explicitly. Thus, to be precise, we shall consider systems of type (2.40) in which U contains two elements u_1 and u_2, $A = A(U)$ is a 2×2 matrix depending explicitly only on U and vector $B \equiv 0$, so that the system becomes

$$U_t + A(U)U_x = 0. \tag{3.1}$$

It will be assumed that the eigenvalues $\lambda^{(1)}, \lambda^{(2)}$ satisfying

$$|A - \lambda I| = 0 \tag{3.2}$$

are real and that the corresponding left eigenvectors $\ell^{(1)}$, $\ell^{(2)}$ satisfying

$$\ell^{(i)}A = \lambda^{(i)}\ell^{(i)} \text{ with } \ell^{(i)} = [\ell_1^{(i)}, \ell_2^{(i)}] \text{ for } i = 1,2 \tag{3.3}$$

span E^2, so that system (3.1) is hyperbolic in the t-direction. Then pre-multiplying equations (3.1) by $\ell^{(i)}$ for i = 1,2 and using relation (3.3) gives as with equation (2.51), the resulting scalar equations

$$\ell^{(1)}(U_t + \lambda^{(1)}U_x) = 0, \tag{3.4a}$$

and

$$\ell^{(2)}(U_t + \lambda^{(2)}U_x) = 0. \tag{3.4b}$$

In these equations $(U_t + \lambda^{(1)}U_x)$ is the directional derivative of U along members of the $C^{(1)}$ family of characteristic curves and $(U_t + \lambda^{(2)}U_x)$ is the corresponding directional derivative along members of the $C^{(2)}$ family. It is at this point that the parametric representation of these curves becomes particularly useful. To distinguish between the $C^{(1)}$ and $C^{(2)}$ families we now parameterise the $C^{(1)}$ family with parameter α and the $C^{(2)}$ family with parameter β, so that the families are determined by integrating

$$C^{(1)} \ : \ \frac{dx}{d\alpha} = \lambda^{(1)}, \quad \frac{dt}{d\alpha} = 1, \tag{3.5a}$$

and

$$C^{(2)} \ : \ \frac{dx}{d\beta} = \lambda^{(2)}, \quad \frac{dt}{d\beta} = 1. \tag{3.5b}$$

Here the parameterisation implied by the second equation in both (3.5a) and (3.5b) is such that α and β are, essentially, the elapsed time measured along respective $C^{(1)}$ and $C^{(2)}$ characteristics relative to reference points on those characteristics. As the α parameterisation varies along different members of the $C^{(1)}$ family it follows that throughout the (x,t)-plane α is a function of both x and t. The situation is strictly analogous for the β parameterisation.

If we require β = const. on members of the $C^{(1)}$ family of characteristics it follows at once that

$$\frac{\partial\beta}{\partial\alpha} = 0 = \frac{\partial\beta}{\partial t}\frac{\partial t}{\partial\alpha} + \frac{\partial\beta}{\partial x}\frac{\partial x}{\partial\alpha} \quad \text{along } C^{(1)} \text{characteristics.} \tag{3.6}$$

As this equation is true along the $C^{(1)}$ family of characteristics, on which now only α varies, the equation may be re-written in the form

$$\frac{\partial\beta}{\partial t}\frac{dt}{d\alpha} + \frac{\partial\beta}{\partial x}\frac{dx}{d\alpha} = 0,$$

which from (3.5a) becomes,

$$\frac{\partial\beta}{\partial t} + \lambda^{(1)}\frac{\partial\beta}{\partial x} = 0. \tag{3.7}$$

Consequently to achieve the desired β parameterisation in the (x,t)-plane it is necessary that $\beta = \beta(x,t)$ should be a solution of equation (3.7). Similarly, if the parameterisation $\alpha = \alpha(x,t)$ in the (x,t)-plane is taken to be such that the corresponding expression for α is true, namely

$$\frac{\partial\alpha}{\partial t} + \lambda^{(2)}\frac{\partial\alpha}{\partial x} = 0, \tag{3.8}$$

then α = const along any member of the $C^{(2)}$ family of characteristic curves. In view of this parameterisation it becomes possible to re-write equations

(3.4a, b) in the more convenient form

$$\ell^{(1)} \frac{dU}{d\alpha} = 0 \quad \text{along } C^{(1)} \text{ characteristics,} \tag{3.9a}$$

and

$$\ell^{(2)} \frac{dU}{d\beta} = 0 \quad \text{along } C^{(2)} \text{ characteristics,} \tag{3.9b}$$

with α and β now in the role of independent variables. The parameterisation itself is illustrated diagramatically in Fig.12 in which representative members of the $C^{(1)}$ and $C^{(2)}$ families of characteristics are indicated.

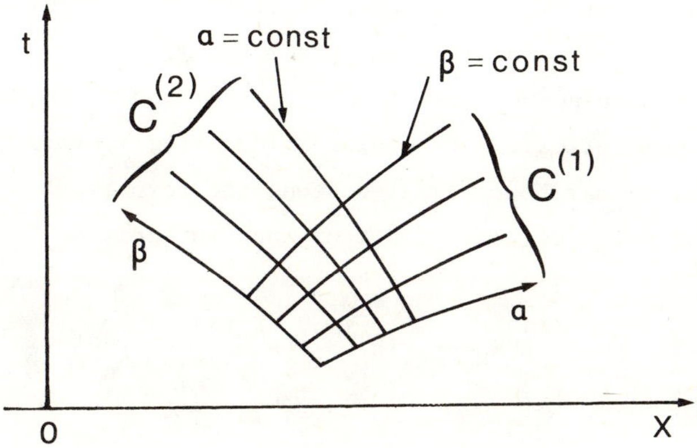

Fig.12 Paramterisation of $C^{(1)}$ and $C^{(2)}$ Families of Characteristics

By adopting the parameters α, β as new independent variables, equations (3.5a, b) and (3.9a, b) may be combined to yield the following properly determined system of four partial differential equations for the variables x, t, u_1, u_2, which then play the role of dependent variables:

$$\frac{dx}{d\alpha} = \lambda^{(1)} \frac{dt}{d\alpha} \quad \text{defining the } C^{(1)} \text{ family,} \tag{3.10a}$$

$$\frac{dx}{d\beta} = \lambda^{(2)} \frac{dt}{d\beta} \quad \text{defining the } C^{(2)} \text{ family,} \tag{3.10b}$$

$$\ell^{(1)} \frac{dU}{d\alpha} = 0, \quad \text{along } C^{(1)} \text{ characteristics,} \tag{3.10c}$$

$$\ell^{(2)} \frac{dU}{d\beta} = 0 \quad \text{along } C^{(2)} \text{ characteristics.} \tag{3.10d}$$

In the literature these equations are often called the characteristic equations for system (3.1).

Solutions of the original system of equations (3.1) will be solutions of the new system (3.10). Furthermore, whenever the Jacobian of the transformation

$$J = \begin{vmatrix} \frac{\partial x}{\partial \alpha} & \frac{\partial x}{\partial \beta} \\ \frac{\partial t}{\partial \alpha} & \frac{\partial t}{\partial \beta} \end{vmatrix} \tag{3.11}$$

is finite and non-singular, it will also be true that a solution of the new system (3.11) is a solution of the old system (3.1).

An immediate consequence of expressing x, t, u_1 and u_2 in terms of α and β is that, after multiplication by scalar integrating factors μ_1 and μ_2, equations (3.10 c, d) may be integrated along the respective $C^{(1)}$ and $C^{(2)}$ families of characteristics. The result of such integration is to give

$$\int \mu_1 \ell_1^{(1)} du_1 + \int \mu_1 \ell_2^{(1)} du_2 = r(\beta) \text{ along } C^{(1)} \text{ characteristics} \tag{3.12a}$$

and

$$\int \mu_2 \ell_1^{(2)} du_1 + \int \mu_2 \ell_2^{(2)} du_2 = s(\alpha) \text{ along } C^{(2)} \text{ characteristics.} \tag{3.12b}$$

The "constants" of integration $r(\beta)$ and $s(\alpha)$ that arise because of the manner of parameterisation of the $C^{(1)}$ and $C^{(2)}$ families are called Riemann invariants. We comment here that for arithmetic convenience some authors introduce a factor 2 on the right hand sides of (3.12 a, b) in the definitions of $r(\beta)$ and $s(\alpha)$. Here $r(\beta)$ will be constant along any particular $C^{(1)}$ characteristic corresponding, say, to $\beta = \beta_1$, though in general it will differ along different characteristics (different β); consequently the Riemann invariant $r(\beta)$ is a function of β defined on the $C^{(1)}$ family of characteristics. The Riemann invariant $s(\alpha)$ has a corresponding interpretation with respect to the $C^{(2)}$ family of characteristics. By virtue of their construction, the Riemann invariants $r(\beta)$, $s(\alpha)$ are functions which depend explicitly only on u_1 and u_2 and which are constant along members of each of their respective families of characteristics. The reason that the dependence of $r(\beta)$ and $s(\alpha)$ on x and t is only implicit, through the functions u_1 and u_2, is on account of the fact that by hypothesis A(U) in system (3.1), and hence

$\ell^{(1)}$, $\ell^{(2)}$, $\lambda^{(1)}$ and $\lambda^{(2)}$, are functions only of U.

We remark here that it was necessary to introduce the scalar integrating factors μ_1, μ_2 because the left eigenvectors $\ell^{(i)} = [\ell_1^{(i)}, \ell_2^{(i)}]$ for i = 1,2 are determinate only up to a multiplicative factor, so that as they stand (3.10 c,d) might not be exact differentials.

It is valuable to interpret equations (3.10c, d) in terms of the (u_1,u_2)-plane. When expanded these equations become

$$\ell_1^{(1)} \frac{du_1}{d\alpha} + \ell_2^{(1)} \frac{du_2}{d\alpha} = 0 \text{ along } C^{(1)} \text{ characteristics,} \tag{3.13a}$$

and

$$\ell_1^{(2)} \frac{du_1}{d\beta} + \ell_2^{(2)} \frac{du_2}{d\beta} = 0 \text{ along } C^{(2)} \text{ characteristics,} \tag{3.13b}$$

where as already noted the coefficients depend explicitly only on u_1 and u_2. Eliminating α and β from equations (3.13) then serves to define two families of curves $\Gamma^{(1)}$ and $\Gamma^{(2)}$ in the (u_1, u_2)-plane which are the solution curves to the following equations

$$\Gamma^{(1)} : \frac{du_1}{du_2} = \zeta^{(1)}, \text{ with } \zeta^{(1)} = -\ell_2^{(1)}/\ell_1^{(1)}, \tag{3.14a}$$

and

$$\Gamma^{(2)} : \frac{du_1}{du_2} = \zeta^{(2)}, \text{ with } \zeta^{(2)} = -\ell_2^{(2)}/\ell_1^{(2)}. \tag{3.14b}$$

As $\ell^{(1)}$, $\ell^{(2)}$ are known functions of u_1 and u_2 the families of curves $\Gamma^{(1)}$, $\Gamma^{(2)}$ may be constructed without reference to a particular solution. In terms of the 2 x 2 matrix $A(U) = \{a_{ij}\}$ for i,j = 1,2, the quantities $\zeta^{(i)}$ in equations (3.14) are easily related to the $\lambda^{(i)}$ by means of equations (3.3). When these are expanded we find

$$\ell_1^{(i)} (a_{11} - \lambda^{(i)}) + \ell_2^{(i)} a_{21} = 0, \tag{3.15a}$$

and

$$\ell_1^{(i)} a_{12} + \ell_2^{(i)} (a_{22} - \lambda^{(i)}) = 0, \text{ for } i = 1,2, \text{ where} \tag{3.15b}$$

$$\lambda^{(1)} = \tfrac{1}{2}\{a_{11} + a_{22} + [(a_{11} + a_{22})^2 - 4(a_{11}a_{22} - a_{12}a_{21})]^{\frac{1}{2}}\}, \tag{3.16a}$$

and

$$\lambda^{(2)} = \tfrac{1}{2}\{a_{11} + a_{22} - [(a_{11} + a_{22})^2 - 4(a_{11}a_{22} - a_{12}a_{21})]^{\frac{1}{2}}\}. \tag{3.16b}$$

Provided $a_{21} \neq 0$ the equation (3.15a) gives

$$\zeta^{(i)} = -\ell_2^{(i)}/\ell_1^{(i)} = (a_{11} - \lambda^{(i)})/a_{21}, \tag{3.17a}$$

whilst if $(a_{22} - \lambda^{(i)}) \neq 0$, the second equation (3.15b) gives

$$\zeta^{(i)} = -\ell_2^{(i)}/\ell_1^{(i)} = a_{12}/(a_{22} - \lambda^{(i)}), \tag{3.17b}$$

either of which may be used to define the $\zeta^{(i)}$ in terms of $\lambda^{(i)}$ for i = 1,2.

If the Jacobian

$$j = \begin{vmatrix} \dfrac{\partial u_1}{\partial x} & \dfrac{\partial u_2}{\partial x} \\ \dfrac{\partial u_1}{\partial t} & \dfrac{\partial u_2}{\partial t} \end{vmatrix} \tag{3.18}$$

is finite and non-singular, equations (3.10 a,b) may be used to determine x and t as functions of u_1 and u_2. More precisely, if u_1 and u_2 are determined in terms of α and β from (3.14 a, b), then either (3.17a) or (3.17b) may be employed to determine the $\lambda^{(i)}$ as functions of α and β. The remaining two variables x and t then follow from (3.10 a, b).

We see from this that by considering the (u_1,u_2)-plane we have interchanged the roles of dependent and independent variables in system (3.1). To pursue this idea a stage further it is necessary to observe that when this interchange is carried out we have, using (3.18), that

$$\frac{\partial u_1}{\partial t} = j\frac{\partial x}{\partial u_2}, \quad \frac{\partial u_2}{\partial t} = -j\frac{\partial x}{\partial u_1} \tag{3.19}$$

$$\frac{\partial u_1}{\partial x} = -j\frac{\partial t}{\partial u_2}, \quad \frac{\partial u_2}{\partial x} = j\frac{\partial t}{\partial u_1} .$$

Provided j is finite and non-singular, substitution of (3.19) into system (3.1) gives rise to the system

$$\begin{bmatrix} 1 & -a_{11} \\ 0 & -a_{21} \end{bmatrix} \begin{bmatrix} x \\ t \end{bmatrix}_{u_2} + \begin{bmatrix} 0 & a_{12} \\ -1 & a_{22} \end{bmatrix} \begin{bmatrix} x \\ t \end{bmatrix}_{u_1} = 0 \tag{3.20}$$

which is seen to be linear in the new dependent variables x and t. This transformation, involving the interchange of dependent and independent variables, has been extensively studied and is called the hodograph transformation [9, 33]. The (u_1, u_2)-plane is then called the hodograph plane. To determine the gradients du_1/du_2 of the characteristic curves in the (u_1, u_2)-plane we now pre-multiply (3.20) by the matrix inverse to the first coefficient matrix to reduce the system to the standard form

$$\begin{bmatrix} x \\ t \end{bmatrix}_{u_2} + \begin{bmatrix} a_{11}/a_{21} & (a_{12}a_{21} - a_{11}a_{22})/a_{21} \\ 1/a_{21} & -a_{22}/a_{21} \end{bmatrix} \begin{bmatrix} x \\ t \end{bmatrix}_{u_1} = 0. \tag{3.21}$$

A simple calculation now verifies that the eigenvalues of the coefficient matrix in system (3.21) are $\zeta^{(1)}$ and $\zeta^{(2)}$ as given in (3.17a, b). The families of curves $\Gamma^{(i)}$ defined in (3.14a, b) are thus the families of characteristic curves of system (3.20) or, equivalently, of system (3.21) and are the images in the hodograph plane of the families of characteristic curves $C^{(i)}$ in the (x,t)-plane. Because of the reduction of system (3.1) to the linear form (3.21) such systems are said to be reducible.

Definition 3.1 (Reducible Systems)
The system

$$U_t + A(U)U_x = 0,$$

in which U is a 2×1 column vector and A(U) is a 2×2 matrix, is said to be reducible if the elements of A(U) depend explicitly only on the elements of U. ▮

In concluding this section let us provide two examples giving rise to Riemann invariants.

1. Unsteady One-Dimensional Isentropic Flow
In one space dimension the system, which has already been encountered in (2.77a, b), becomes

$$\begin{bmatrix} \rho \\ u \end{bmatrix}_t + \begin{bmatrix} u & \rho \\ a^2/\rho & u \end{bmatrix} \begin{bmatrix} \rho \\ u \end{bmatrix}_x = 0, \tag{3.22}$$

which is seen to be reducible according to Definition 3.1. The eigenvalues $\lambda^{(i)}$ and corresponding left eigenvectors $\ell^{(i)}$ of

$$A = \begin{bmatrix} u & \rho \\ a^2/\rho & u \end{bmatrix}$$

are

$$\lambda^{(1)} = u + a, \; \lambda^{(2)} = u - a, \; \ell^{(1)} = [a/\rho,\ 1], \; \ell^{(2)} = [a/\rho,\ -1]. \quad (3.23)$$

When these are substituted into (3.12 a, b) and the identifications $u_1 \to \rho$, $u_2 \to u$ are made it is apparent that no integrating factors are required, for we obtain

$$\int \frac{a}{\rho}\, d\rho + u = r(\beta) \text{ along } C^{(1)} \text{ characteristics} \quad (3.24a)$$

and

$$\int \frac{a}{\rho}\, d\rho - u = s(\alpha) \text{ along } C^{(2)} \text{ characteristics,} \quad (3.24b)$$

where the characteristics themselves are determined as the solutions to

$$C^{(i)} : \frac{dx}{dt} = \lambda^{(i)}, \text{ for } i = 1,2. \quad (3.24c)$$

For a polytropic gas in which $p = K\rho^{\gamma}$, $a^2 = dp/d\rho$ so that (3.24a, b) may be integrated to give

$$\frac{2a}{\gamma - 1} + u = r(\beta) \text{ and } \frac{2a}{\gamma-1} - u = s(\alpha) \quad (3.25a)$$

or, equivalently,

$$u = \tfrac{1}{2}(r-s) \text{ and } a = \tfrac{1}{4}(\gamma-1)(r+s). \quad (3.25b)$$

So, in this case, the Riemann invariants $r(\beta)$, $s(\alpha)$ are given by (3.24 a, b). Had the left eigenvectors $[1, \pm \rho/a]$ been used, which differ from those in (3.23) by the factor a/ρ, then integrating factors would have been required in (3.12a, b). In fact it would have been necessary to set $\mu_1 = \mu_2 = a/\rho$ to obtain (3.24 a,b).

2. Shallow Water Wave Approximation

This system given in (1.17) becomes reducible when the seabed is flat, for then $H_x \equiv 0$. The eigenvalues $\lambda^{(i)}$ and left eigenvectors $\ell^{(i)}$ are

$$\lambda^{(1)} = u + c, \quad \lambda^{(2)} = u - c, \quad \ell^{(1)} = [1,2], \quad \ell^{(2)} = [1,-2].$$

Considering the case when $H_x \equiv 0$ and making the identifications $u_1 \to u$, $u_2 \to c$ in (3.12 a,b) we find that the Riemann invariants are simply

$$u + 2c = r(\beta) \quad \text{along } C^{(1)} \text{ characteristics} \tag{3.26a}$$

and

$$u - 2c = s(\alpha) \quad \text{along } C^{(2)} \text{ characteristics.} \tag{3.26b}$$

As usual, the characteristics themselves are determined as the solutions to

$$C^{(i)} : \frac{dx}{dt} = \lambda^{(i)} \quad \text{for } i = 1,2. \tag{3.26c}$$

When expressed in terms of r, s the variables u, c become

$$u = \tfrac{1}{2}(r + s), \quad c = \tfrac{1}{4}(r - s). \tag{3.27}$$

To follow through the concept of reducibility in this case we re-write (3.26c) in parametric form as

$$C^{(1)} : \frac{dx}{d\alpha} = (u + c)\frac{dt}{d\alpha} \tag{3.28a}$$

$$C^{(2)} : \frac{dx}{d\beta} = (u - c)\frac{dt}{d\beta} \tag{3.28b}$$

and use the simple form of (3.27) together with the fact that $r = r(\beta)$ and $s = s(\alpha)$ to replace α in (3.28a) by s and β in (3.28b) by r when in the (r,s)-plane we obtain

$$C^{(1)} : \frac{\partial x}{\partial s} = \tfrac{1}{4}(3r + s)\frac{\partial t}{\partial s}, \tag{3.29a}$$

and

$$C^{(2)} : \frac{\partial x}{\partial r} = \tfrac{1}{4}(r + 3s)\frac{\partial t}{\partial r}. \tag{3.29b}$$

Equating the mixed derivatives $\partial^2 x/\partial s\partial r$ and $\partial^2 x/\partial r\partial s$ then gives the linear second order equation

$$2(r - s)\frac{\partial^2 t}{\partial r\partial s} + 3\frac{\partial t}{\partial s} - 3\frac{\partial t}{\partial r} = 0. \tag{3.30}$$

The solution to the shallow water wave approximation for water above a flat seabed then follows by solving (3.30) for t and then using equations

(3.29a,b) to determine x. Equation (3.30) is of the Euler-Poisson-Darboux (E.P.D) type and for more information we refer to the basic paper by Darboux [47] and to the review by Weinstein [48]. A comprehensive survey of the application of the hodograph method to gas dynamics has been given by Manwell [33] whilst Bers [49] analyses transonic gas flow using both the hodograph transformation and other methods.

3.2 Simple Waves

In general, the simplification brought about by the linearity of (3.20) in the (u_1,u_2)-plane which enables the curves $\Gamma^{(i)}$ to be constructed without reference to a particular solution is balanced, and possibly nullified, by the complexity of the task of determining the corresponding curves $C^{(i)}$ in the (x,t)-plane. However there is one related and important class of problems in which this difficulty does not occur and this gives rise to the class of so called simple wave solutions. These occur when the Jacobian $j \equiv 0$, so that the hodograph formulation (3.20) of the problem becomes invalid. We can, however, revert to equations (3.10), which are still true, even though it is not then possible to use equations (3.10a,b) to construct the solution when u_1 and u_2 have been determined.

Let us examine this situation more closely by finding under what conditions j may vanish. The simplest case is when $u_1(x,t) = u_{01}$(const.) and $u_2(x,t) = u_{02}$(const.) in some region $\mathscr{B}$ of the (x,t)-plane. This entire region $\mathscr{B}$ then maps to the single point B at (u_{01},u_{02}) in the (u_1,u_2)-plane and, by analogy with physical problems, $\mathscr{B}$ will be called a region of constant state. The characteristics $C^{(i)}$ in the (x,t)-plane will then be families of straight lines and the general situation is illustrated in Fig. 13.

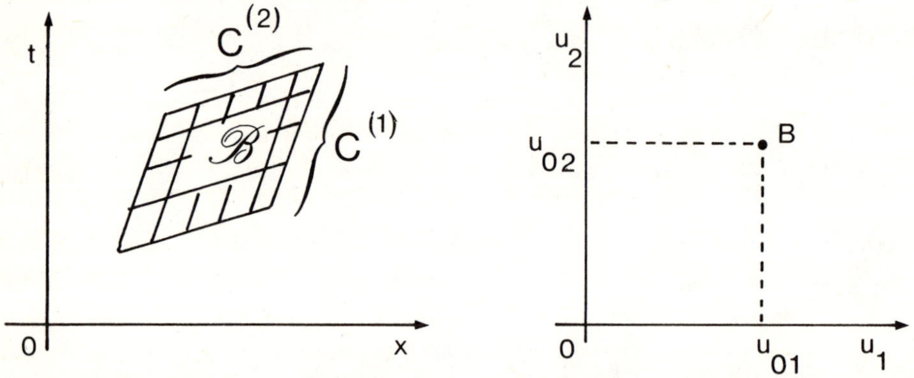

(a) Region of constant state $\mathscr{B}$

(b) Region of constant state $\mathscr{B}$ maps to the single point B at (u_{01},u_{02}).

Fig. 13. Constant state region.

The Jacobian j will also vanish if u_1 and u_2 are not independent by virtue of the fact that $u_2 = f(u_1)$ in some region $\mathscr{S}$ of the (x,t)-plane, where f is some differentiable function. The situation in the (u_1,u_2)-plane is then that the entire region $\mathscr{S}$ maps to some curve in the (u_1,u_2)-plane. A straightforward substitution of $u_2 = f(u_1)$ into system (3.1) verifies that the consequence of this functional dependence is that the curve $u_2 = f(u_1)$ must lie along arcs constructed from the $\Gamma^{(1)}$ and $\Gamma^{(2)}$ families of characteristics. It is precisely this last type of situation that gives rise to simple wave solutions. The region $\mathscr{S}$ in the (x,t)-plane in which $u_2 = f(u_1)$ is then called a simple wave region.

Let us suppose that a simple wave region $\mathscr{S}$ has as its image in the (u_1,u_2)-plane the characteristic $\Gamma^{*(2)}$, then because of the mapping of characteristics $C^{(i)}$ onto characteristics $\Gamma^{(i)}$, all characteristics $C^{(2)}$ traversing $\mathscr{S}$ must have $\Gamma^{*(2)}$ as their image. Each characteristic in the simple wave region then has as its image a point P on $\Gamma^{*(2)}$. Consequently all along the characteristic $C_P^{(1)}$ that maps to the point P on $\Gamma^{*(2)}$ in the (u_1,u_2)-plane with coordinates (u_{1P},u_{2P}) it must follow that $u_1 = u_{1P}$ and $u_2 = u_{2P}$. Thus u_1 and u_2 are constant along each of the $C^{(1)}$ characteristics in the simple wave region. It may thus be concluded from (3.10a) that in the simple wave region the $C^{(1)}$ characteristics must be straight lines. As different characteristics correspond to different points on $\Gamma^{*(2)}$, each straight line $C^{(1)}$ characteristic will have a different slope. Conversely, if the simple wave region $\mathscr{S}$ has as its image the characteristic $\Gamma^{*(1)}$, then it will be the $C^{(2)}$ characteristics in $\mathscr{S}$ that become straight lines.

In the class of continuous solutions, the fact that a constant state region maps to a single point in the (u_1,u_2)-plane together with the fact just established that a simple wave region is traversed by a family of straight line characteristics may be combined to yield an important result. It is that if u_1,u_2 are constant along an arc of a characteristic $C_Q^{(1)}$ then the regions adjacent to $C_Q^{(1)}$ will either be a constant state region or a simple wave region. This property is illustrated in Fig. 14. The region of constant state $\mathscr{B}$ in Fig. 14(a) maps to point Q in Fig. 14(b) whilst the simple wave region $\mathscr{S}$ in Fig. 14(a) maps to the arc $\Gamma^{*(2)}$ in Fig. 14(b). The restriction to continuous solutions is important because, as we shall see later, when discontinuous solutions are permitted simple wave solutions are no longer appropriate. Continuity of the solution across the characteristic $C_Q^{(1)}$

bounding the region of constant state ensures coincidence with $C_Q^{(1)}$ of the characteristic bounding the simple wave region $\mathscr{S}$. The previous results may be formulated as a theorem in the following manner.

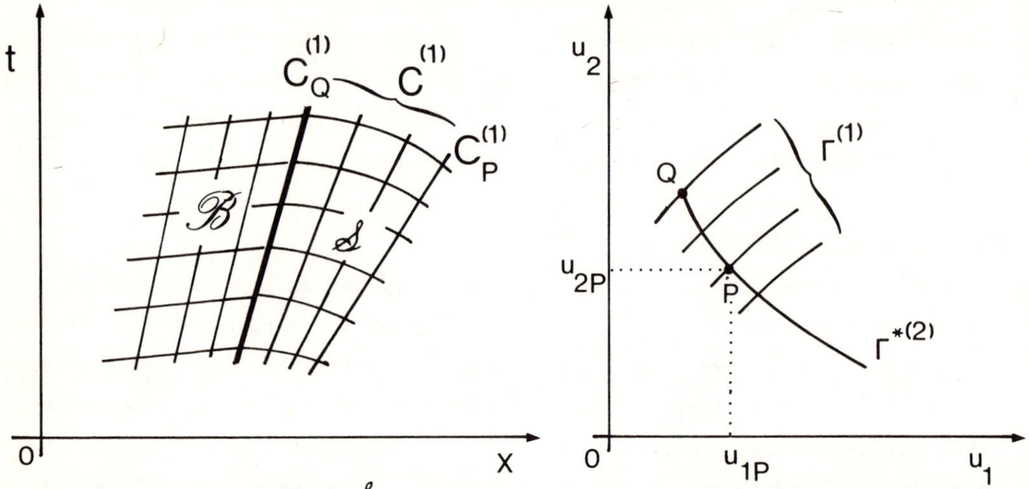

(a) Simple wave region $\mathscr{S}$ adjacent to a region of constant state.

(b) Image in (u_1,u_2)-plane of simple wave region adjacent to a region of constant state.

Fig.14. Simple wave adjacent to constant state.

Theorem 3.1 (Simple Wave Regions)

Let the system

$$U_t + A(U)U_x = 0 \tag{3.31}$$

be reducible, where U is a 2 X 1 column vector and A(U) is a 2 X 2 matrix. Then in a simple wave region $\mathscr{S}$ associated with (3.31) it follows that:

a) one of the two families of characteristic curves traversing $\mathscr{S}$ will always be a family of straight lines;

b) the family of straight line characteristics in the simple wave region has as its image in the (u_1,u_2) - plane a single arc Γ;

c) the variables u_1 and u_2 are constant along members of a family of straight line characteristics traversing $\mathscr{S}$;

d) if u_1 and u_2 are constant along a characteristic, then in the regions adjacent to that characteristic the solution is either a constant state solution or a simple wave;

e) within the class of continuous solutions the solution adjacent to a region of constant state is a simple wave. ▮

As a simple wave region is characterised by there being a functional relationship of the form $u_2 = f(u_1)$ between u_1 and u_2, it is natural that when data is given for a simple wave along an arc $\mathscr{A}$ in the (x,t)-plane that it should reflect this fact. Accordingly, if the arc $\mathscr{A}$ is parameterised in terms of the variable σ by writing $x = x(\sigma)$, $t = t(\sigma)$, then at points $(x(\sigma), t(\sigma))$ of $\mathscr{A}$ we shall assign u_1 and u_2 by setting $u_1 = u_1(\sigma)$, $u_2 = u_2(\sigma)$ in such a way that the elimination of σ between these two functions gives the required relationship $u_2 = f(u_1)$ along $\mathscr{A}$. If the $C^{(1)}$ family of characteristics is the straight line family associated with the simple wave region, then the gradient of the $C^{(1)}$ characteristic through $(x(\sigma), t(\sigma))$ is $\lambda^{(1)}(u_1(\sigma), u_2(\sigma))$. The characteristic field may be parameterised in terms of ξ by writing

$$\frac{dx}{d\xi} = \lambda^{(1)}(\sigma), \qquad \frac{dt}{d\xi} = 1$$

from which the solution in terms of the parameters σ, ξ follows by integration:

$$x = x(\sigma) + \xi\lambda^{(1)}(\sigma), \quad t = t(\sigma) + \xi \tag{3.32a}$$

with

$$u_1 = u_1(\sigma), \quad u_2 = u_2(\sigma). \tag{3.32b}$$

When the $C^{(1)}$ characteristics diverge the simple wave $\mathscr{S}$ characterises an expansion, and when they converge it characterises compression. These situations are illustrated in Fig. 15a and Fig. 15b, respectively, and in the compression case it is apparent that the solution ceases to be unique after the envelope $\mathscr{E}$ has formed. This matter will be taken up again in the next chapter but we mention here the work by Smoller andJohnson [50] in connection with the Riemann problem, which involves assigning piecewise constant boundary data to the arc $\mathscr{A}$, and the related paper by Smoller [51] concerning the uniqueness of the solution to the Riemann problem. The third and last case to be mentioned is that of the so-called centred simple wave. This is illustrated in Fig. 15c and it occurs when the arc $\mathscr{A}$ degenerates to a single point A. Clearly the solution has a singularity at point A but it is well defined in the simple wave region $\mathscr{S}$ traversed by the $C^{(1)}$ family of straight line characteristics.

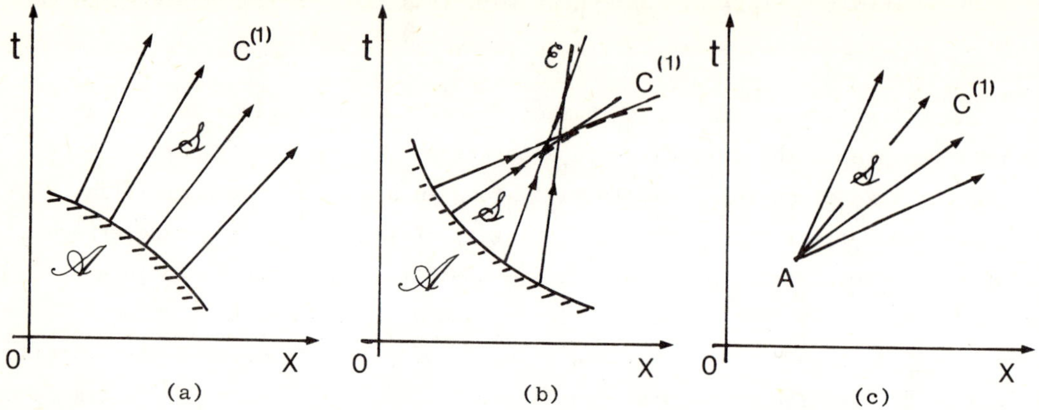

Expansion simple wave

Compression simple wave and envelope formation

Centred simple wave

Fig.15. Expansion, Compression and Centred Simple Waves.

The application of simple waves to gas dynamics has been described in detail by Courant and Friedrichs [9], Von Mises [52] and Rozhdestvensky and Yanenko [53]. An account of their application to magnetohydrodynamics has been given by Jeffrey and Taniuti [8] and to nonlinear continuum mechanics by Eringen and Suhubi [54]. An interesting application to chromatography has also been described by Rhee, Aris and Amundson [55].

3.3 Generalised Simple Waves

It is reasonable to enquire whether the notion of a simple wave can be generalised and extended to systems with more than two dependent variables. Specifically we shall consider homogeneous systems in one space dimension and time which involve vectors U with n components ($n \geq 2$) and $n \times n$ matrices A(U) with elements that depend explicitly only on U; that is systems

$$\begin{bmatrix} u_1 \\ u_2 \\ \cdot \\ \cdot \\ \cdot \\ u_n \end{bmatrix}_t + \begin{bmatrix} a_{11}(U) & a_{12}(U) & \dots & a_{1n}(U) \\ a_{21}(U) & a_{22}(U) & \dots & a_{2n}(U) \\ \cdot & \cdot & \cdots & \cdot \\ a_{n1}(U) & a_{n2}(U) & \dots & a_{nn}(U) \end{bmatrix} \begin{bmatrix} u_1 \\ u_2 \\ \cdot \\ \cdot \\ \cdot \\ u_n \end{bmatrix}_x = 0, \qquad (3.33)$$

which will be said to be reducible in the generalised sense. In what follows it is not necessary that system (3.33) should be of conservation form.

Before developing the main theme of this section we employ a simple algebraic argument to show that a direct extension of the concept of a Riemann invariant is not possible when $n > 2$. If possible, let us suppose there to be a function $J(u_1, u_2, \ldots, u_n)$ associated with a family of characteristics, say the $C^{(k)}$ family, with the property that J = const. along each characteristic belonging to that family.

Then J must satisfy the condition

$$\frac{\partial J}{\partial u_1} du_1 + \frac{\partial J}{\partial u_2} du_2 + \ldots + \frac{\partial J}{\partial u_n} du_n = 0 \text{ along each } C^{(k)}$$

or, as $du_j = (\partial u_j/\partial x)dx + (\partial u_j/\partial t)dt$,

$$\sum_{j=1}^{n} \frac{\partial J}{\partial u_j}\left(\frac{\partial u_j}{\partial t} + \frac{\partial u_j}{\partial x}\frac{dx}{dt}\right) = 0 \quad \text{along each } C^{(k)}. \tag{3.34}$$

Since $dx/dt = \lambda^{(k)}$ along the $C^{(k)}$ family of characteristics, (3.34) becomes the following condition on J

$$(\nabla_u J)\frac{dU}{dk} = 0 \quad \text{along each } C^{(k)}, \tag{3.35}$$

where $\nabla_u \equiv \{\partial/\partial u_1, \partial/\partial u_2, \ldots, \partial/\partial u_n\}$ is the gradient operator with respect to $(u_1, u_2, \ldots, u_n)$-space, and

$$\frac{d}{dk} \equiv \frac{\partial}{\partial t} + \lambda^{(k)}\frac{\partial}{\partial x}.$$

Comparison of (3.35) with the equation that results after pre-multiplying (3.33) by the k-th left eigenvector $\ell^{(k)}$ establishes that $(\nabla_u J)$ must be proportional to $\ell^{(k)}$. As this vector is only determined up to an arbitrary multiplicative factor, the result $(\nabla_u J) \propto \ell^{(k)}$ imposes $n-1$ conditions on the n elements of $(\nabla_u J)$.

There also exist other constraints amongst the elements of $(\nabla_u J)$, because the equality of mixed derivatives requires that

$$\frac{\partial}{\partial u_i}\left(\frac{\partial J}{\partial u_j}\right) = \frac{\partial}{\partial u_j}\left(\frac{\partial J}{\partial u_i}\right) \quad \text{for } i, j = 1, 2, \ldots, n. \tag{3.36}$$

Of these n^2 conditions the n corresponding to $i = j$ are identities, and so must be discounted, whilst from the remaining $n^2 - n$ conditions, symmetry arguments show that there are in fact only $\frac{1}{2}(n^2-n)$ different conditions. Thus the n components of $(\nabla_u J)$ are required to satisfy $N = \frac{1}{2}(n-1)(n+2)$ conditions, comprising the n-1 from $(\nabla_u J) \propto \ell^{(k)}$ and the $\frac{1}{2}n(n-1)$ from (3.36). In general this is not possible unless n=2, so that it is apparent that some weaker form of extension of the notion of a Riemann invariant must be employed

Since our concern will be with generalised simple waves we seek the desired weaker extension of a Riemann invariant from amongst the properties of ordinary simple waves. The property we choose to generalise is that in an ordinary simple wave there is a functional dependence between u_1 and u_2 of the form $u_1 = f(u_2)$. Accordingly, we propose to take a generalised simple wave region to be one in which the solution vector U is a function of only one of its n-elements, say of u_1, so that $U = U(u_1)$. This is the definition adopted by Jeffrey and Taniuti [8], and as we show later it is equivalent to the one proposed by Lax [56].

If in (3.33) we set $U = U(u_1)$, so that $u_i = u_i(u_1)$ for i=2,3,...,n, an elementary calculation establishes that

$$\left(\frac{\partial u_1}{\partial t} I + \frac{\partial u_1}{\partial x} A\right)\frac{dU}{du_1} = 0. \tag{3.37}$$

This system can only have a non-trivial solution if

$$|A - \mu I| = 0, \tag{3.38}$$

where $\mu = -(\partial u_1/\partial t)/(\partial u_1/\partial x)$. The n solutions $\mu^{(i)}$ to the algebraic equation (3.38) are just the eigenvalues $\lambda^{(i)}$ of A, so that when $\mu = \mu^{(i)}$ the vector dU/du_1 must be proportional to the right eigenvector $r^{(i)}$ of A corresponding to $\lambda^{(i)}$. As system (3.33) is assumed to be hyperbolic there will be n distinct eigenvectors $r^{(i)}$. The fact that $\mu^{(i)} = \lambda^{(i)}$ then implies that along the family of characteristic curves $C^{(i)}$

$$\frac{dx}{dt} = \lambda^{(i)} = -\left(\frac{\partial u_1}{\partial t}\right)\bigg/\left(\frac{\partial u_1}{\partial x}\right) \tag{3.39}$$

or, equivalently, that

$$\frac{\partial u_1}{\partial x}\,dx + \frac{\partial u_1}{\partial t}\,dt = 0 \text{ along each } C^{(i)},$$

thereby showing that

$$u_1(x,t) = \text{const along each } C^{(i)}. \tag{3.40}$$

A corresponding result applies along each of the n different families of characteristics $C^{(i)}$.

Consider now the k-th such family and let us determine the nature of the generalised simple wave that is associated with it. The fact that the proposed generalisation of a simple wave region allows a corresponding generalisation of the notion of Riemann invariants will emerge from the fact that we will find that we are able to determine the form of these generalised invariants.

Setting i = k in (3.40) shows that u_1 = const along the $C^{(k)}$ family. This fact, taken together with $U = U(u_1)$, then shows U = const along members of the k-th family of characteristics. As $A = A(U)$ and U = const along any $C^{(k)}$ characteristic we conclude that $\lambda^{(k)}$ = const, thereby proving that the $C^{(k)}$ family comprises a family of straight line characteristics. Now provided attention is confined to continuous and differentiable solutions, system (3.37) may be written in differential form by replacing dU/du_1 by dU, when the fact that in the $C^{(k)}$ family of characteristics dU/du_1 is proportional to the right eigenvector $r^{(k)}$ establishes that $dU \propto r^{(k)}$ along each member of the family of straight line characteristics $C^{(k)}$.

This result gives rise to the set of n differential equations

$$\frac{du_1}{r_1^{(k)}} = \frac{du_2}{r_2^{(k)}} = \ldots = \frac{du_n}{r_n^{(k)}}, \tag{3.41}$$

in which $r_1^{(k)}, r_2^{(k)}, \ldots, r_n^{(k)}$ are the elements of the eigenvector $r^{(k)}$. These n first order ordinary differential equations determine the behaviour of the solution U across what will be called a generalised $\lambda^{(k)}$- simple wave. When integrated, (3.41) will give rise to n-1 linearly independent relations between the n elements of U, though multiplication of (3.41) by an integrating factor $m(u_1,u_2,\ldots,u_n)$ might be necessary because of the fact that $r^{(k)}$ is only determined up to an arbitrary multiplicative factor.

These n-1 invariant relations along the k-th family of characteristics will be denoted by

$$J_i^{(k)}(U) = \text{const for } i = 1,2,\ldots,n-1. \tag{3.42}$$

They will be called generalised $\lambda^{(k)}$-Riemann invariants to make clear that they are associated with the k-th family of characteristics $C^{(k)}$. These relations hold throughout the generalised $\lambda^{(k)}$-simple wave region where they determine the behaviour of a continuous and differentiable solution.

On occasions, when deriving the generalised $\lambda^{(k)}$-Riemann invariants from (3.41), it is useful to express them in terms of a parameter ξ by writing (3.41) in the form

$$\frac{du_1}{r_1^{(k)}} = \frac{du_2}{r_2^{(k)}} = \ldots = \frac{du_n}{r_n^{(k)}} = d\xi. \tag{3.43}$$

The u_i may then be determined in terms of ξ by integrating the system

$$\frac{du_j}{d\xi} = r_j^{(k)}, \quad \text{for } j=1,2,\ldots,n. \tag{3.44}$$

Elimination of ξ between these n equations gives rise to the n-1 generalised $\lambda^{(k)}$-Riemann invariants (3.42).

We are now in a position to give a formal definition of a generalised $\lambda^{(k)}$-simple wave, though we preface it by an extended definition of reducibility.

Definition 3.2 (Reducible System in Generalised Sense)
The system

$$U_t + AU_x = 0$$

in which U is an $n \times 1$ column vector and A is an $n \times n$ matrix will be said to be reducible in the generalised sense if the elements of A depend explicitly only on U. ∎

Definition 3.3 (Generalised Simple Wave Region)
Let the system

$$U_t + AU_x = 0$$

in which U is an $n \times 1$ column vector and A is an $n \times n$ matrix be reducible in the generalised sense. Then any region $\mathcal{S}$ in the (x,t)-plane in which the solution vector U is of the form $U = U(u_j)$, with u_j one particular element of U, will be called a generalised simple wave region. ∎

It is now useful to interpret generalised $\lambda^{(k)}$-simple waves in terms of $(u_1,u_2,\ldots,u_n)$-space, just as was done in Section 3.2 for ordinary simple waves. The fact that the family of $C^{(k)}$ characteristics in a generalised $\lambda^{(k)}$-simple wave are straight lines which transport constant values of U means that in the class of differentiable solutions with which we work each $C^{(k)}$ characteristic maps to a point of an arc Γ in the $(u_1,u_2,\ldots,u_n)$-space. This arc is then the image in the $(u_1,u_2,\ldots,u_n)$-space of the entire generalised $\lambda^{(k)}$-simple wave region in the (x,t)-space. As in the ordinary simple wave case, a region of constant state adjacent to a generalised $\lambda^{(k)}$-simple wave region will map to a single point at the end of the arc Γ. Here the term region of constant state is used to signify a region in which the solution is constant. The families of characteristic curves associated with the other n-1 eigenvalues of A will have as their images segments of the arc Γ. In terms of the generalised $\lambda^{(k)}$-Riemann invariants $J_i^{(k)}$, the arc Γ is simply the line of intersection in the $(u_1,u_2,\ldots,u_n)$-space of the (n-1)-dimensional manifolds $J_i^{(k)}$ = const, for i=1,2,...,n-1.

The results established so far can be given formal expression as the following theorem.

Theorem 3.2 (Generalised Simple Wave Regions)

Let the system

$$U_t + AU_x = 0$$

with U an $n \times 1$ column vector and A an $n \times n$ matrix be reducible in the generalised sense. Then if $\mathcal{S}$ is a generalised simple wave region:

a) there is a family of straight line characteristics $C^{(k)}$ traversing $\mathcal{S}$;

b) the solution vector U is constant along members of the $C^{(k)}$ family;

c) in $\mathcal{S}$ there will be n-1 generalised $\lambda^{(k)}$-Riemann invariants $J_i^{(k)}(U)$ = const for i = 1,2,...,n-1 which will be determined by integrating equations (3.41). ▌

Various other generalisations of simple waves and Riemann invariants are also possible, some more useful than others. For an extension to problems involving acceleration fronts in general elastic materials in $\mathbb{R}^3 \times t$ we refer to Varley [57], while these same ideas have been applied by Parker [58] to a study of steady supersonic flow over developable surfaces. See also the paper by Cumberbatch and Varley [59]. For a very abstract approach to generalised simple waves and Riemann invariants we mention the work by Peradzynski [60] and by Burnat [61]. Applications to the ideas of the present section to continuum mechanics are also to be found in the paper by Collins [14] and in the paper by Jeffrey and Teymur [62] who were concnerned with wave front propagation and shock formation.

The interesting and difficult interaction problem for waves also makes appeal to Riemann invariants, and studies for the case of reducible systems in gas dynamics and other disciplines are to be found in Courant and Friedrich [9], Whitham [7] and von Mises [52], while for the equivalent problems in chemically reacting systems we mention the paper by Rhee, Aris and Amundson [55]. For wave interactions involving elasticity and gas dynamics that necessitate the use of generalised Riemann invariants reference should be made to Collins [14] and to Rozhdestvensky and Yanenko [53], and for magnetohydrodynamics to Jeffrey and Taniuti [8].

The following example provides the details of the determination of generalised $\lambda^{(k)}$-Riemann invariants in non-isentropic gas flow and, as a corollary, it also establishes that simple wave flow in a polytropic gas must, in fact, be isentropic.

<u>One-dimensional Unsteady Non-isentropic Flow</u>

The system of equations in a polytropic gas [9] is

$$U_t + AU_x = 0, \tag{3.45a}$$

with

$$U = \begin{bmatrix} \rho \\ u \\ S \end{bmatrix} \quad \text{and } A = \begin{bmatrix} u & \rho & 0 \\ (\partial p/\partial \rho)/\rho & u & (\partial p/\partial S)/\rho \\ 0 & 0 & u \end{bmatrix}, \tag{3.45b}$$

where ρ is the gas density, u is the gas velocity in the x-direction and $S = S(p,\rho)$ is the entropy, with $p = p(\rho,S)$ the gas pressure and $a^2 = \partial p/\partial \rho$ the square of the sound speed.

The eigenvalues $\lambda^{(i)}$ of A are

$$\lambda^{(1)} = u + a, \quad \lambda^{(2)} = u - a, \quad \lambda^{(3)} = u, \tag{3.46a}$$

while the corresponding right eigenvectors $r^{(i)}$ are

$$r^{(1)} = \begin{bmatrix} 1 \\ a/\rho \\ 0 \end{bmatrix}, \quad r^{(2)} = \begin{bmatrix} 1 \\ -a/\rho \\ 0 \end{bmatrix}, \quad r^{(3)} = \begin{bmatrix} -(\partial p/\partial S) \\ 0 \\ (\partial p/\partial \rho) \end{bmatrix}. \tag{3.46b}$$

Inserting these results into (3.44) and combining the resulting equations to eliminate ξ then gives:

<u>Generalised $\lambda^{(1)}$-Riemann invariants</u>

$$J_1^{(1)} \equiv u - \int \frac{a}{\rho}\, d\rho = \text{const}, \text{ and } J_2^{(1)} \equiv S = \text{const} \tag{3.47a}$$

<u>Generalised $\lambda^{(2)}$-Riemann invariants</u>

$$J_1^{(2)} \equiv u + \int \frac{a}{\rho}\, d\rho = \text{const}, \text{ and } J_2^{(2)} \equiv S = \text{const} \tag{3.47b}$$

<u>Generalised $\lambda^{(3)}$-Riemann invariants</u>

$$J_1^{(3)} \equiv p = \text{const}, \text{ and } J_2^{(3)} \equiv u = \text{const}. \tag{3.47c}$$

This last result has an interesting consequence in respect of the assumed non-isentropic nature of the gas flow in the generalised $\lambda^{(k)}$-simple wave regions. Consider the generalised $\lambda^{(3)}$-Riemann invariants. Then as $S = S(\rho,p)$ and $p = \text{const}$ we have $S = \tilde{S}(\rho)$, while since $p = p(\rho,S)$ and $S = \tilde{S}(\rho)$ we have $p = \tilde{p}(\rho)$. However $p = \text{const}$, so that $\rho = \text{const}$, and hence $S = \text{const}$. As S is then constant in all three $\lambda^{(k)}$-Riemann invariants it follows that the flow must be isentropic whenever it takes place in the associated generalised simple wave region. The third equation in system (3.45) is thus redundant, because it is identically zero. The fact that generalised simple wave flow must be isentropic, which follows trivially from the use of generalised Riemann invariants, was first established for multidimensional gas flow by Levine [63] who proved the fact by means of a different form of argument.

3.4. Exceptional Condition and Genuine Nonlinearity

It is now appropriate to introduce two related concepts in connection with first order quasilinear hyperbolic systems. These are the notions of a solution which is exceptional with respect to a particular characteristic field, and of a system which exhibits genuine nonlinearity with respect to a characteristic field. Although these ideas may be introduced without reference to generalised simple waves it will be convenient to use this approach here and to remove this restriction later.

For our starting point we take a generalised $\lambda^{(k)}$-simple wave region and the associated generalised $\lambda^{(k)}$-Riemann invariants $J_i^{(k)}(U) = \text{const}$ for $i=1,2,\ldots,n-1$. Each of these invariants defines a manifold in the $(u_1, u_2, \ldots, u_n)$-space, on the i-th of which $J_i^{(k)}$ must obey the constraint condition $dJ_i^{(k)} = 0$, or

$$\frac{\partial J_i^{(k)}}{\partial u_1} du_1 + \frac{\partial J_i^{(k)}}{\partial u_2} du_2 + \ldots + \frac{\partial J_i^{(k)}}{\partial u_n} du_n = 0. \tag{3.48}$$

Now in a generalised $\lambda^{(k)}$-simple wave region we have from (3.44) that $du_j = r_j^{(k)} d\xi$, so that (3.48) is equivalent to the condition

$$(\nabla_u J_i^{(k)}) r^{(k)} = 0, \tag{3.49}$$

with $i = 1,2,\ldots,n-1$. These orthogonality conditions for the $(\nabla_u J_i^{(k)})$ with respect to the right eigenvector $r^{(k)}$ associated with eigenvalue $\lambda^{(k)}$ of A were the ones used by Lax [56] to define generalised $\lambda^{(k)}$-Riemann invariants. He then used this definition to establish the properties of solutions in a generalised simple wave region that are given in our Theorem 3.2.

Before proceeding to our main objective let us first use condition (3.49), together with an argument due to Friedrichs [64] (see also [56]), to prove that the solution adjacent to a region of constant state must be a generalised simple wave region. This result which might have been conjectured from Theorem 3.2 will then complement the results of that theorem.

First we notice that from Theorem 3.2 it follows that if a region $\mathscr{A}$ of constant state exists in the (x,t)-plane, then it will be bounded by a characteristic, say by a member C of the $C^{(k)}$-family. Any region $\mathscr{S}$ adjacent to it will also be bounded by this same line C.

Now pre-multiplication of system (3.1) by the left eigenvector $\ell^{(j)}$ of A gives the system

$$\ell^{(j)} \frac{dU}{dj} = 0 \qquad \text{along the } C^{(j)} \text{ family}, \tag{3.50}$$

for $j = 1,2,\ldots,n$. As the left and right eigenvectors of A are biorthogonal, so that

$$\ell^{(j)} r^{(k)} = 0 \qquad \text{for } j \neq k,$$

it follows directly from (3.49) that the vector $\ell^{(j)}$ must be expressible as a linear combination of the vectors $(\nabla_u J_i^{(k)})$. Accordingly, we set

$$\ell^{(j)} = \sum_{s=1}^{n-1} b_{js}(\nabla_u J_s^{(k)}) \text{ for } j \neq k. \tag{3.51}$$

Equations (3.50) then become

$$\sum_{s=1}^{n-1} b_{js}(\nabla_u J_s^{(k)})\frac{dU}{dj} = 0 \quad \text{for } j \neq k, \tag{3.52}$$

which by the chain rule reduces to

$$\sum_{s=1}^{n-1} b_{js} \frac{dJ_s^{(k)}}{dj} = 0 \quad \text{for } j \neq k. \tag{3.53}$$

This is now a linear hyperbolic system involving (n-1) equations for the (n-1) generalised $\lambda^{(k)}$-Riemann invariants $J_1^{(k)}, J_2^{(k)}, \ldots, J_{n-1}^{(k)}$. For any specified solution vector U the coefficients b_{js} will be known. The condition $j \neq k$ ensures that the line C common to both the region of constant state and the generalised simple wave region will not be a characteristic of the new system. Consequently there exists a unique smooth solution that can be continued across the line C. Since the solution on one side of C was the constant state solution, the solution that is continued across it will be one for which all the generalised $\lambda^{(k)}$-Riemann invariants are constant. Hence from the nature of generalised Riemann invariants it may be seen that the solution adjacent to a region of constant state must be a generalised simple wave region. This result also merits a formal statement.

Theorem 3.3 (Constant State and Generalised Simple Wave)
Let the system

$$U_t + AU_x = 0$$

with U an $n \times 1$ column vector and A an $n \times n$ matrix be reducible in the generalised sense. Then if $\mathscr{A}$ is a region of constant state in the (x,t)-plane, the region $\mathscr{S}$ adjacent to it is a generalised simple wave region. ▮

Let us now examine further the implications of equation (3.49). Consider the special case in which the eigenvalue $\lambda^{(k)}$ is expressible as a function of

$J_1^{(k)}, J_2^{(k)}, \ldots, J_{n-1}^{(k)}$. Then we have

$$(\nabla_u \lambda^{(k)}) = \left\{ \sum_{m=1}^{n-1} \frac{\partial \lambda^{(k)}}{\partial J_m^{(k)}} \frac{\partial J_m^{(k)}}{\partial u_1}, \sum_{m=1}^{n-1} \frac{\partial \lambda^{(k)}}{\partial J_m^{(k)}} \frac{\partial J_m^{(k)}}{\partial u_2}, \ldots, \sum_{m=1}^{n-1} \frac{\partial \lambda^{(k)}}{\partial J_m^{(k)}} \frac{\partial J_m^{(k)}}{\partial u_n} \right\}$$

or, equivalently,

$$(\nabla_u \lambda^{(k)}) = \sum_{m=1}^{n-1} \frac{\partial \lambda^{(k)}}{\partial J_m^{(k)}} \nabla_u J_m^{(k)}, \tag{3.54}$$

showing that $(\nabla_u \lambda^{(k)})$ is a linear combination of the gradients of the generalised $\lambda^{(k)}$-Riemann invariants. After post-multiplication of (3.54) by $r^{(k)}$ it then follows directly from (3.49) that

$$(\nabla_u \lambda^{(k)}).r^{(k)} = 0. \tag{3.55}$$

In general, when a quasilinear hyperbolic system exists for which property (3.55) is true with respect to the k-th characteristic field $C^{(k)}$ associated with $\lambda = \lambda^{(k)}$, the system will be said to be exceptional with respect to the k-th characteristic field. This will be true irrespective of whether or not the system permits generalised simple wave solutions. When (3.55) is not true, the system of equations will be said to be genuinely nonlinear with respect to the k-th characteristic field $C^{(k)}$. Expressed differently, condition (3.55) asserts that when a system is exceptional with respect to the k-th characteristic field, the directional derivative of $\lambda^{(k)}$ in the direction of the eigenvector $r^{(k)}$ is zero. We now formulate these ideas generally, without reference to generalised simple waves or to Riemann invariants.

Definition 3.4 (Exceptional Condition and Genuine Nonlinearity) Consider the quasilinear hyperbolic system

$$U_t + AU_x + B = 0,$$

where U is an $n \times 1$ column vector, $A = A(U,x,t)$ is an $n \times n$ matrix and

$B = B(U,x,t)$ is an $n \times 1$ column vector. Then the system will be said to be:

a) exceptional with respect to the k-th characteristic field if

$$(\nabla_u \lambda^{(k)})r^{(k)} = 0;$$

b) completely exceptional if it is exceptional with respect to each of the n characteristic fields corresponding to $\lambda^{(1)}, \lambda^{(2)}, \ldots, \lambda^{(n)}$;

c) genuinely nonlinear with respect to the k-th characteristic field if

$$(\nabla_u \lambda^{(k)})r^{(k)} \neq 0. \quad \blacksquare$$

General One-Dimensional Unsteady Non-isentropic Flow

The exceptional condition and also the concept of genuine nonlinearity are illustrated by equations (3.45). For them $\nabla_u \equiv \{\partial/\partial\rho, \partial/\partial u, \partial/\partial S\}$, so that it follows from (3.46a) that:

$$(\nabla_u \lambda^{(1)}) = (\partial a/\partial\rho, \quad 1 \quad, \quad 0),$$

$$(\nabla_u \lambda^{(2)}) = (-\partial a/\partial\rho, \quad 1 \quad, \quad 0),$$

$$(\nabla_u \lambda^{(3)}) = (\quad 0 \quad, \quad 1 \quad, \quad 0).$$

Using these results with (3.46b) to form the products $(\nabla_u \lambda^{(k)})r^{(k)}$ then shows

$$(\nabla_u \lambda^{(1)})r^{(1)} = \frac{\partial a}{\partial\rho} + \frac{a}{\rho} \neq 0,$$

$$(\nabla_u \lambda^{(2)})r^{(2)} = -\left(\frac{\partial a}{\partial\rho} + \frac{a}{\rho}\right) \neq 0,$$

$$(\nabla_u \lambda^{(3)})r^{(3)} = 0.$$

Consequently system (3.45) is genuinely nonlinear with respect to the $\lambda^{(1)}$ and $\lambda^{(2)}$ characteristic fields, but it is exceptional with respect to the $\lambda^{(3)}$ characteristic field. This last fact is, of course, to be expected when generalised simple wave flow is involved because of the conclusions reached at the end of Section 3.3.

Non-Physical Example Showing Completely Exceptional Behaviour

The system (1.19a,b) was shown on page 56 to be totally hyperbolic and to have the eigenvalues $\lambda^{(1)} = -1$, $\lambda^{(2)} = 0$ and $\lambda^{(3)} = 1$. As $(\nabla_u \lambda^{(k)}) \equiv 0$ for $k = 1,2,3$ it follows immediately that this system is completely exceptional. We shall consider this example further in Section 3.6.

3.5 Evolution of Discontinuities in Solutions from Arbitrary Initial Data

The purpose of the present section will be to examine further the evolution of non-differentiable discontinuous solutions from arbitrary initial data. Some preliminary consideration of this matter was given in Section 1.6 for a scalar equation, but we now extend the examination to reducible systems involving two equations. The analysis, which is based on a paper by Jeffrey [46], will be formulated in terms of the Riemann invariant representation of the solution introduced in Section 3.1 for reducible systems. It will be seen that the genuinely nonlinear condition introduced in Definition 3.4 appears in the final result in a natural manner.

Related investigations have been carried out by Lax [65], who also examined reducible systems, and also by John [66], who used genuine nonlinearity and related ideas as tools in a penetrating examination of the formation of singularities in one-dimensional nonlinear wave propagation involving a general system of n equations. Certain aspects of the problem examined by John will form the subject matter of Chapter 5, though there the analysis will proceed in a different manner from that employed by John.

Our starting point will be system (3.1), that is to say a quasilinear hyperbolic reducible system

$$U_t + A(U)U_x = 0, \tag{3.56a}$$

in which

$$U = \begin{bmatrix} u_1 \\ u_2 \end{bmatrix} \quad \text{and} \quad A = \begin{bmatrix} a_{11} & a_{12} \\ a_{21} & a_{22} \end{bmatrix}, \tag{3.56b}$$

with $a_{ij} = a_{ij}(u_1, u_2)$. Then, proceeding as in Section 3.1, we arrive at the pair of equations (3.12a,b) which, for convenience, we repeat below:

$$\int \mu_1 \ell_1^{(1)} du_1 + \int \mu_1 \ell_2^{(1)} du_2 = r(\beta) \text{ along } C^{(1)} \text{ characteristics} \tag{3.57a}$$

$$\int \mu_2 \ell_1^{(2)} du_1 + \int \mu_2 \ell_2^{(2)} du_2 = s(\alpha) \text{ along } C^{(2)} \text{ characteristics.} \tag{3.57b}$$

The $C^{(1)}$ and $C^{(2)}$ families of characteristics are defined, as usual, as the solution curves to the differential equations

$$C^{(i)} : \quad \frac{dx}{dt} = \lambda^{(i)} \text{ for } i = 1,2, \tag{3.58}$$

with $\lambda^{(i)}$ the eigenvalues of A.

When the Jacobian of the transformation from the (u_1, u_2)-plane to the (r,s)-plane is finite and non-singular, it follows that the characteristic equations (3.10c,d) that led to (3.57a,b) (i.e. to (3.12a,b)) may be written

$$\frac{dr}{d\alpha} = \frac{\partial r}{\partial t} + \lambda^{(1)} \frac{\partial r}{\partial x} = 0 \text{ along } C^{(1)} \text{ characteristics} \tag{3.59a}$$

and

$$\frac{ds}{d\beta} = \frac{\partial s}{\partial t} + \lambda^{(2)} \frac{\partial s}{\partial x} = 0 \text{ along } C^{(2)} \text{ characteristics} \tag{3.59b}$$.

For convenience of reference the α, β parameterisation employed here is indicated again in Fig.16.

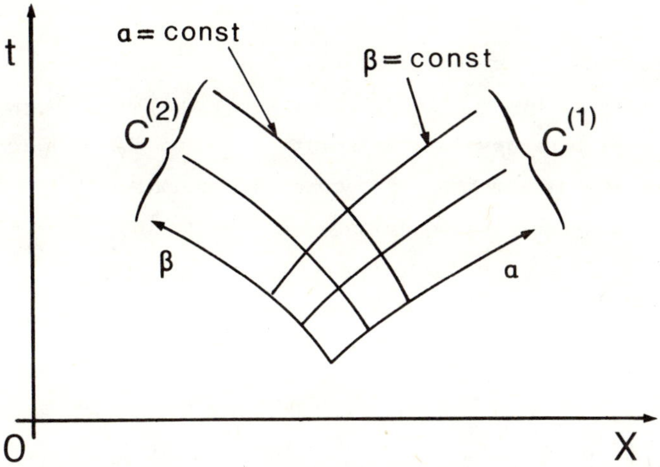

Fig.16 The α, β-parameterisation.

The argument that now follows proceeds in three distinct steps. The first involves the derivation of the differential equation satisfied by a Riemann invariant along the family of characteristics along which it varies. The second is the derivation of a comparison theorem that is required, together with the necessary uniqueness theorem for the solution. The third and final step is the derivation of asymptotic estimates for the time of breakdown of the differentiable solution.

(i) Differential equations along characteristics The expression $dr/d\alpha = 0$ in equation (3.59a) involves the behaviour of r_t and r_x along $C^{(1)}$ characteristics, so we now use it to determine the differential equations satisfied by these functions along those same $C^{(1)}$ characteristics. This will be achieved by deriving two new equations from (3.59a) by differentiation with respect to α and to β, yielding

$$\left(\frac{\partial}{\partial t} + \lambda^{(1)}\frac{\partial}{\partial x}\right)\left(\frac{\partial r}{\partial t} + \lambda^{(1)}\frac{\partial}{\partial x}\right) = 0 \text{ and}$$

$$\left(\frac{\partial}{\partial t} + \lambda^{(2)}\frac{\partial}{\partial x}\right)\left(\frac{\partial r}{\partial t} + \lambda^{(1)}\frac{\partial r}{\partial x}\right) = 0, \tag{3.60}$$

both of which are valid along $C^{(1)}$ characteristics. As the system (3.56) is assumed to be totally hyperbolic, so that $\lambda^{(1)} \neq \lambda^{(2)}$, it follows immediately that

$$\frac{\partial}{\partial t}\left(\frac{\partial r}{\partial t} + \lambda^{(1)}\frac{\partial r}{\partial x}\right) = 0 \quad \text{and} \quad \frac{\partial}{\partial x}\left(\frac{\partial r}{\partial t} + \lambda^{(1)}\frac{\partial r}{\partial x}\right) = 0 \tag{3.61}$$

along $C^{(1)}$ characteristics. The fact that $\lambda^{(1)}$ and $\lambda^{(2)}$ are finite, and $\partial r/\partial t$ and $\partial r/\partial x$ are related by equation (3.59a), means that it will be sufficie to determine the time of breakdown of the solution due to the Riemann invariar r by determining the time at which $\partial r/\partial x$ becomes unbounded.

We are thus led to consider the equation

$$\frac{\partial}{\partial x}\left(\frac{\partial r}{\partial t} + \lambda^{(1)}\frac{\partial r}{\partial x}\right) = 0 \quad \text{along} \quad C^{(1)} \text{ characteristics.} \tag{3.62}$$

Performing the indicated differentiaton gives

$$\frac{\partial^2 r}{\partial t\partial x} + \frac{\partial\lambda^{(1)}}{\partial r}\left(\frac{\partial r}{\partial x}\right)^2 + \frac{\partial\lambda^{(1)}}{\partial s}\left(\frac{\partial s}{\partial x}\right)\left(\frac{\partial r}{\partial x}\right) + \lambda^{(1)}\frac{\partial^2 r}{\partial x^2} = 0, \tag{3.63}$$

which must hold along $C^{(1)}$ characteristics. A similar result holds along $C^{(2)}$ characteristics, for which strictly analogous reasoning shows that the equatic

$$\frac{\partial^2 s}{\partial t\partial x} + \frac{\partial\lambda^{(2)}}{\partial s}\left(\frac{\partial s}{\partial x}\right)^2 + \frac{\partial\lambda^{(2)}}{\partial r}\left(\frac{\partial r}{\partial x}\right)\left(\frac{\partial s}{\partial x}\right) + \lambda^{(2)}\frac{\partial^2 s}{\partial x^2} = 0 \tag{3.64}$$

describes the variation of $\partial s/\partial x$. This second equation then determines the time of breakdown of the solution due to the Riemann invariant s.

Returning to equation (3.63) we now simplify it by first recognising that the first and last terms combine to form $d/d\alpha\,(\partial r/\partial x)$. Then, applying the operator identity $d/d\alpha \equiv d/d\beta + (\lambda^{(1)} - \lambda^{(2)})\partial/\partial x$ to $s(\alpha)$, we see that $ds/d\alpha = (\lambda^{(1)} - \lambda^{(2)})(\partial s/\partial x)$, so that equation (3.63) reduces to

$$\frac{d}{d\alpha}\left(\frac{\partial r}{\partial x}\right) + \frac{\partial\lambda^{(1)}}{\partial r}\left(\frac{\partial r}{\partial x}\right)^2 + \frac{1}{(\lambda^{(1)} - \lambda^{(2)})}\left(\frac{ds}{d\alpha}\right)\left(\frac{\partial\lambda^{(1)}}{\partial s}\right)\left(\frac{\partial r}{\partial x}\right) = 0. \tag{3.65}$$

This homogeneous nonlinear ordinary differential equation along the $C^{(1)}$ characteristics can be further simplified if the dependent variable is changed so that only a quadratic term in the new dependent variable remains. To achieve this we write $v_1 = (\partial r/\partial x)f_1$, thereby reducing (3.65) to

$$\frac{dv_1}{d\alpha} + \frac{1}{f_1}\left(\frac{\partial\lambda^{(1)}}{\partial r}\right)v_1^{\,2} = 0, \tag{3.66}$$

provided the function f_1 is chosen such that it satisfies the equation

$$\frac{1}{f_1}\frac{df_1}{ds} = \left(\frac{1}{\lambda^{(1)} - \lambda^{(2)}}\right)\frac{\partial\lambda^{(1)}}{\partial s} . \tag{3.67}$$

Thus f_1 is given by

$$f_1 = K \exp\left\{\int\left(\frac{1}{\lambda^{(1)} - \lambda^{(2)}}\right)\left(\frac{\partial\lambda^{(1)}}{\partial s}\right) ds\right\} , \tag{3.68}$$

where K is an arbitrary constant which, for convenience, we take to be unity. So, defining g_1 by the relation

$$g_1 = \int\left(\frac{1}{\lambda^{(1)} - \lambda^{(2)}}\right)\left(\frac{\partial\lambda^{(1)}}{\partial s}\right) ds, \tag{3.69}$$

equation (3.66) determining the variation of the new dependent variable v_1 along the $C^{(1)}$ characteristics takes the form

$$\frac{dv_1}{d\alpha} + \left(\frac{\partial\lambda^{(1)}}{\partial r}\right)\left[\exp\{-g_1\}\right]v_1^{\,2} = 0. \tag{3.70}$$

Identical reasoning shows that if $v_2 = (\partial s/\partial x)f_2$, where

$$f_2 = K' \exp\left\{\int\left(\frac{1}{\lambda^{(2)} - \lambda^{(1)}}\right)\left(\frac{\partial\lambda^{(2)}}{\partial r}\right) dr\right\} , \tag{3.71}$$

and g_2 is defined in a manner analogous to g_1, then v_2 satisfies the equation

$$\frac{dv_2}{d\beta} + \left(\frac{\partial\lambda^{(2)}}{\partial s}\right)\left[\exp\{-g_2\}\right]v_2^{\,2} = 0, \qquad (3.72)$$

along $C^{(2)}$ characteristics. Equations (3.70) and (3.72) will be taken as the starting point for the derivation of the asymptotic estimates of the critical time t_c at which the solution of the system of equations (3.59) first develops a singularity due either to v_1 or v_2 becoming unbounded along their respective characteristics.

(ii) Uniqueness and comparison theorems. To pose a pure initial value problem for the system of equations (3.59) it is necessary to specify the behaviour of the column vector U on the initial line $t = 0$. Accordingly we set

$$U(x,0) = \Phi(x), \qquad (3.73)$$

where $\Phi(x)$ is a 2×1 column vector whose elements depend only on x, and suppos that the elements of $\Phi(x)$ are at least Lipschitz continuous on the x-axis. Then, by virtue of equations (3.57a,b), the vector $\Phi(x)$ determines the values of r and s which are to be associated with the $C^{(1)}$ and $C^{(2)}$ characteristics passing through each point of the x-axis. As the coefficients of the matrix A(U) are assumed to be at least continuous and piecewise differentiable functions of u_1 and u_2, this implies that the coefficients of $v_1^{\,2}$ and $v_2^{\,2}$ in equations (3.70) and (3.72) must be at least Lipschitz continuous.

Now both equations (3.70) and (3.72) are of the form

$$\frac{dw}{dx} = A(x,w), \qquad (3.74)$$

with $A(x,w)$ Lipschitz continuous with respect to w in an appropriate region of the (x,w)-plane. Since by a well known result [67, Chapter 1, Theorem 2.2] the solution w of equation (3.75) is unique, it immediately follows that the solutions v_1 and v_2 of equations (3.70) and (3.72) are unique.

Now the nonlinear variable coefficient equations (3.70) and (3.72) determining the behaviour of v_1 and v_2 along the $C^{(1)}$ and $C^{(2)}$ characteristics cannot be integrated in the form shown. To overcome this difficulty we obtain asymptotic results by applying a useful comparison theorem which we now establish. To see the nature of the problem that is involved we begin by considering the form of equation (3.70) that is appropriate to a $C^{(1)}$ characteristic through the point $x = \xi$ on the initial line $t = 0$. The coefficient

$(\partial\lambda^{(1)}/\partial r)\exp\{-g_1\}$ in this equation is a function of α and $v_1(\alpha)$ once the point $x = \xi$ on the initial line through which the defining $C^{(1)}$ characteristic passes and the initial vector $\phi(x)$ have been specified.

Let us denote this coefficient by $-A(\alpha,\xi)$, so that the constant value $-A_o(\xi)$ assumed by it when $\alpha = \alpha_o$, say, at the point $x = \xi$ on the initial line will be determined by the relation $A_o(\xi) = A(\alpha_o,\xi)$. Let us now compare the solutions v_1 and $\bar{v}_1$ of the two equations

$$\frac{dv_1}{d\alpha} = A(\alpha,\xi)v_1^2 \tag{3.75}$$

and

$$\frac{d\bar{v}_1}{d\alpha} = A_o(\xi)\bar{v}_1^2, \tag{3.76}$$

respectively, where both v_1 and $\bar{v}_1$ are subject to some initial conditions that are to be specified later and $A(\alpha,\xi)$ is Lipschitz continuous with respect to α. The comparison solution $\bar{v}_1$ is thus the solution of a constant coefficient equation.

To simplify the argument introduce the Lipschitz continuous function $B(\alpha,v_1(\alpha),\xi) = A(\alpha,\xi)v_1^2$ with the Lipschitz constant $K(\xi)$, and the continuous function $B_o(\bar{v}_1(\alpha),\xi) = A_o(\xi)\bar{v}_1^2$. Assume further that they satisfy the condition

$$|B(\alpha,w,\xi) - B_o(w,\xi)| \le M(\xi), \tag{3.77}$$

for some range of α, say, $\alpha_o \leqq \alpha \leqq \alpha_1$, where α_1 is chosen such that w remains uniformly bounded for all values of $A(\alpha,\xi)$ and $A_o(\xi)$ appropriate to all points ξ on the initial line. That a choice of such an α_1 is possible will be established later.

The final forms of the equations to be examined are thus

$$\frac{dv_1}{d\alpha} = B(\alpha,v_1(\alpha),\xi) \tag{3.78}$$

and

$$\frac{d\bar{v}_1}{d\alpha} = B_o(\bar{v}_1(\alpha),\xi). \tag{3.79}$$

Applying an obvious extension of a standard comparison theorem [67, Chapter 1, Theorem 2.1] it follows at once that

$$|v_1(\alpha) - \bar{v}_1(\alpha)| \le |v_1(\alpha_o) - \bar{v}_1(\alpha_o)| \exp\{K(\xi)\,|\alpha - \alpha_o|\} + \frac{M(\xi)}{K(\xi)}[\exp\{K(\xi)\,|\alpha - \alpha_o|\} - 1]. \tag{3.80}$$

It is natural to adopt the same initial conditions for the comparison solution $\bar{v}_1$ as for the genuine solution v_1. Thus, setting $v_1(\alpha_o) = \bar{v}_1(\alpha_o)$ on the initial lines gives the inequality

$$|v_1(\alpha) - \bar{v}_1(\alpha)| \le \frac{M(\xi)}{K(\xi)}[\exp\{K(\xi)\,|\alpha - \alpha_o|\} - 1]. \tag{3.81}$$

This result furnishes a bound for the modulus of the error between the required solution v_1 and the comparison solution $\bar{v}_1$ at a point with coordinate α on the $C^{(1)}$ characteristic through $x = \xi$ on the initial line. A similar result is also valid for the solution v_2 of equation (3.72) and its related comparison solution $\bar{v}_2$.

Later the comparison equation (3.76) will be integrated, so that when the form of $(\partial\lambda/\partial r)\exp\{-g_1\}$ is specified and $K(\xi)$ and $M(\xi)$ are known, inequality (3.81) may be used to estimate the modulus of its difference from v_1 for $\alpha_o \le \alpha \le \alpha_1$. The discussion of the estimate for v_2 is strictly analogous to that for v_1 and so it now only remains to show the existence of the number α_1 introduced in connection with condition (3.77). This will be accomplished in the following sub-section, prior to determining the asymptotic estimate of the critical time.

(iii) <u>Asymptotic estimate of the critical time t_c</u>. To determine the behaviour of the comparison solution $\bar{v}_1$ as a function of the time t, it is necessary to recall that the parameterisation α is such that it is essentially the elapsed time t along the $C^{(1)}$ characteristics. The parameter β has a similar meaning along the $C^{(2)}$ characteristics. Direct integration of constant coefficient equation (3.76) then shows that

$$\bar{v}_1(t) = \frac{\bar{v}_{01}(\xi)}{1 - tA_o(\xi)\bar{v}_{01}(\xi)}, \tag{3.82}$$

where $\bar{v}_{01}(\xi)$ denotes the initial value of $\bar{v}_1$ at the point $x = \xi$ on the initial line $t = 0$ through which the characteristic passes on which $\bar{v}_1(t)$ is defined.

This expression shows that when $A_o(\xi)\bar{v}_o(\xi)$ is positive, the comparison solution $\bar{v}_1$ becomes unbounded on the defining $C^{(1)}$ characteristics at the elapsed critical times $T_c^{(1)}(A_o(\xi),\xi)$ determined by the expression

$$T_c^{(1)}(A_o(\xi),\xi) = \frac{1}{A_o(\xi)\bar{v}_{o1}(\xi)} . \tag{3.83}$$

The superscript 1 and the argument ξ are used here to signify that $T_c^{(1)}(A_o(\xi),\xi)$ is the comparison solution critical time on the $C^{(1)}$ characteristic that passes through the point with coordinate $x = \xi$ on the initial line $t = 0$.

Similar reasoning shows that the comparison solution $\bar{v}_2$ becomes unbounded on the defining $C^{(2)}$ characteristics whenever a positive critical time is determined by the expression

$$T_c^{(2)}(\bar{A}_o(\eta),\eta) = \frac{1}{\bar{A}_o(\eta)\bar{v}_{o2}(\eta)} . \tag{3.84}$$

The superscript 2 signifies that $T_c^{(2)}(\bar{A}_o(\eta),\eta)$ is the comparison solution critical time on the $C^{(2)}$ characteristic passing through the point with coordinate $x = \eta$ on the initial line $t = 0$. In this expression $\bar{A}_o(\eta)$ has been used to denote the constant coefficient in the comparison equation for $\bar{v}_2$ that corresponds to the coefficient $A_o(\xi)$ in equation (3.76).

Provided the derivatives $\partial\lambda^{(1)}/\partial r$ and $\partial\lambda^{(2)}/\partial s$ do not change sign along the initial line, then by choosing the signs of the Riemann invariants so that these derivatives are negative, the coefficients $A_o(\xi)$ and $\bar{A}_o(\eta)$ may always be taken as positive, thereby ensuring that the elapsed critical times given by (3.83) and (3.84) are positive whenever $\bar{v}_{o1}$, $\bar{v}_{o2}$ are positive. This is equivalent to adjusting the signs of r and s by choosing the signs of the integrating factors μ_1 and μ_2 in (3.57a,b) in an appropriate manner.

Let us now establish the existence of the number α_1 used in condition (3.77) and examine the relationship of the comparison solution critical time $T_c^{(1)}(A(\xi),\xi)$ to the critical time $t_c^{(1)}(\xi)$ that is appropriate to the actual solution v_1. Here again the result for v_2 is strictly analogous and so only the details of the argument for the variable v_1 will be presented.

Consider first equation (3.76) defined along the $C^{(1)}$ characteristic through the point $x = \xi$ on the initial line (hereafter replacing α by t), together with the new comparison equation

$$\frac{d\tilde{v}_1}{dt} = N\tilde{v}_1^{\,2}, \tag{3.85}$$

in which N is some positive constant. Take $\tilde{v}_1 = v_1$ when $t = 0$, so that equations (3.75) and (3.85) have identical initial conditions.

If equation (3.75) is subtracted from this equation the following result is easily established

$$\frac{d}{dt}[\tilde{v}_1 - v_1] = [N - A(t,\xi)]\tilde{v}_1^{\,2} + A(t,\xi)[\tilde{v}_1^{\,2} - v_1^{\,2}].$$

Using the initial condition $v_1 = \tilde{v}_1$ when $t = 0$ shows that the rate and direction of change of the difference $(\tilde{v}_1 - v_1)$ is determined by the magnitude and sign of $N - A(t,\xi)$. When, for example, $A(t,\xi) > 0$ and it is possible to choose a positive N dependent on ξ, which we shall write as $N(\xi)$, such that $N(\xi) - A(t,\xi) > 0$ throughout the range of integration with respect to time t, we may at once conclude that $v_1 < \tilde{v}_1$, so that $\tilde{v}_1$ then provides an upper bound for v_1 on the defining characteristic. Since we have seen from equation (3.83) that for a comparison equation of this kind $\tilde{v}_1$ will cease to be defined beyond the time

$$T_c^{(1)}(N(\xi),\xi) = \frac{1}{N(\xi)\tilde{v}_{01}(\xi)},$$

it immediately follows that, along the characteristic that is involved, $T_c^{(1)}(N(\xi),\xi)$ provides a lower bound for the time for which the exact solution v_1 exists.

Provided $A(t,\xi)$ does not change sign along a characteristic (i.e., $\partial\lambda^{(1)}/\partial r$ and $\partial\lambda^{(2)}/\partial s$ remain negative everywhere), by taking $N(\xi)$ to be the least upper bound $n(\xi)$ of $A(t,\xi)$ along the characteristic we obtain the expression $T_{inf}^{(1)}(\xi) = 1/n(\xi)\tilde{v}_{01}(\xi)$ for the greatest lower bound of the times $T_c^{(1)}(N(\xi),\xi)$. This is the best estimate that may be obtained by this method of the lower bound for the critical time for the genuine solution v_1 associated with the $C^{(1)}$ characteristic through $x = \xi$ on the initial line.

To obtain the analytic form of $T_{inf}^{(1)}(\xi)$ we now need to employ the defining relations

$$N(\xi) = -\left(\frac{\partial\lambda^{(1)}}{\partial r}\right)\exp\{-g_1(r,s)\}, \quad \tilde{v}_{01}(\xi) = \left[\left(\frac{\partial r}{\partial x}\right)\exp\{g_1(r,s)\}\right]_{t=0}.$$

It will be recalled that $x = \xi$ on the initial line determines a particular $C^{(1)}$ characteristic, and so corresponds to some definite value of r which we shall denote by $r_o(\xi)$. However s, which varies along the $C^{(1)}$ characteristic, is some function of time with the initial value $s = s_o(\xi)$ at $t = 0$, since on the initial line $r(x,t)$ and $s(x,t)$ are of the form $r_o(x) = r(x,0)$ and $s_o(x) = s(x,0)$. The expression for $T_{inf}^{(1)}(\xi)$ thus becomes

$$T_{inf}^{(1)}(\xi) = \frac{-1}{\left(\frac{\partial r}{\partial x}\right)_{\substack{t=0\\x=\xi}} \exp\{g_1(r_o(\xi), s_o(\xi))\} \sup\left[\left(\frac{\partial \lambda^{(1)}}{\partial r}\right)\exp\{-g_1(r_o(\xi),s)\}\right]_s}, \tag{3.86}$$

where $\sup\,[\cdot]_s$ denotes the supremum with respect to s, the variable r_o being held constant.

Each $C^{(1)}$ characteristic has such a time $T_{inf}^{(1)}(\xi)$ associated with it, and if the greatest lower bound of $T_{inf}^{(1)}(\xi)$ is denoted by ${}_*T^{(1)}$, then ${}_*T^{(1)}$ is the desired lower bound of the times for which the comparison solution exists.

The critical time ${}_*T^{(1)}$ associated with the $C^{(1)}$ characteristics thus has the form

$${}_*T^{(1)} = \inf\left[\frac{-1}{\left(\frac{\partial r}{\partial x}\right)_{\substack{t=0\\x=\xi}} \exp\{g_1(r_o(\xi), s_o(\xi))\} \sup\left[\left(\frac{\partial \lambda^{(1)}}{\partial r}\right)\exp\{-g_1(r_o(\xi),s)\}\right]_s}\right]_\xi, \tag{3.87}$$

and is the least upper bound of the numbers α_1 involved in inequality (3.77) as ξ ranges over the initial line. That this bound exists follows immediately from the fact that for a properly posed problem $\Phi(x)$ and the initial Riemann invariant distributions $r_o(x) = r(x,0)$ and $s_o(x) = s(x,0)$ are continuous and finitely bounded functions so that $T_{inf}^{(1)}(\xi)$, which is a continuous function of them, must itself be finitely bounded. The time ${}_*T^{(1)}$ must be positive, so that a breakdown in the solution due to the $C^{(1)}$ characteristics will occur whenever

$$\left(\frac{\partial r}{\partial x}\right)_{\substack{t=0\\x=\xi}} > 0.$$

This establishes the existence of a number $\alpha_1 < {}_*T^{(1)}$ with the property that the function w in the inequality (3.77) remains uniformly bounded for all points ξ on the initial line provided $\alpha_o < \alpha < \alpha_1$.

A similar argument when applied to the $C^{(2)}$ characteristics gives rise to a critical time ${}_*T^{(2)}$ associated with the $C^{(2)}$ characteristics of the form

$$ {}_*T^{(2)} = \inf\left[\frac{-1}{\left(\frac{\partial s}{\partial x}\right)_{\substack{t=0\\x=\eta}} \exp\{g_2(r_o(\eta),s_o(\eta))\} \sup\left[\left(\frac{\partial\lambda^{(1)}}{\partial s}\right)\exp\{-g_2(r,s_o(\eta))\}\right]_r}\right]_\eta . \tag{3.88} $$

Since both the $C^{(1)}$ and $C^{(2)}$ characteristics can give rise to breakdown of the solution, it finally follows that the number t_{inf}, defined to be the least positive number of the pair ${}_*T^{(1)}$ and ${}_*T^{(2)}$, provides the lower bound for the time of existence of a solution of the comparison equations. Since $\tilde{v}_i$ majorizes v_i it follows also that t_{inf} is the best estimate obtainable by this method of the lower bound for the time of existence of a solution of the original system of equations (3.56).

Returning to equation (3.85), and this time identifying the number N dependent on ξ with the number $N'(\xi)$, which we now suppose satisfies the inequality $N'(\xi) - A(t,\xi) < 0$, enables the previous argument to be used to establish that $\tilde{v}_1 < v_1$. Thus on the defining $C^{(1)}$ characteristic the corresponding time $T_c^{(1)}(N'(\xi),\xi)$ provides an upper bound to the time of existence of a solution for $\tilde{v}_1$ and thus also for v_1. Continuing the argument in an analogous fashion then leads to the determination of a number t_{sup} which is the best estimate obtainable by this method of the upper bound for the time of existence of a solution of the original system of equations (3.56). The number t_{sup} is thus the least positive number of the two quantities

$$ {}^*T^{(1)} = \inf\left[\frac{-1}{\left(\frac{\partial r}{\partial x}\right)_{\substack{t=0\\x=\xi}} \exp\{g_1(r_o(\xi),s_o(\xi))\} \inf\left[\left(\frac{\partial\lambda^{(1)}}{\partial r}\right)\exp\{-g_1(r_o(\xi),s)\}\right]_s}\right]_\xi , \tag{3.89} $$

for the estimate along $C^{(1)}$ characteristics, and the corresponding expression for the estimate along the $C^{(2)}$ characteristics,

$$*T^{(2)} = \inf\left[\frac{-1}{\left(\frac{\partial s}{\partial x}\right)_{\substack{t=0\\x=\eta}} \exp\{g_2(r_o(\eta), s_o(\eta))\} \inf\left[\left(\frac{\partial\lambda^{(2)}}{\partial s}\right)\exp\{-g_2(r, s_o(\eta))\}\right]_r}\right]_\eta \tag{3.90}$$

The actual value t_c of the time of existence of a solution of the original system of equations (3.56) then satisfies the inequality

$$t_{inf} < t_c < t_{sup}. \tag{3.91}$$

Here the numbers t_{inf} and t_{sup} are to be interpreted in the sense that the solution is certainly bounded for $t < t_{inf}$, whilst the solution is certainly unbounded for $t > t_{sup}$.

Normally the bounds (3.91) are wide apart, but under certain conditions the difference $(t_{sup} - t_{inf})$ becomes small, so that a good estimate of t_c may be obtained. This can happen, for example, when the Riemann invariant distributions $r_o(x)$ and $s_o(x)$ differ only slightly from the constant values $\hat{r}_o$ and $\hat{s}_o$, respectively. Under these conditions, since r and s are constant along respective characteristics, they also can differ only slightly from $\hat{r}_o$ and $\hat{s}_o$ at all subsequent times along the characteristics until breakdown of the solution occurs. In addition to this, continuous functions of r and s can vary only slightly from constant values. Consequently, the expressions occurring in (3.87) to (3.90) which define t_{inf} and t_{sup} may thus be approximated by the much simpler expressions:

t_{inf} is the least positive number of the two quantities

$$\frac{-1}{\max\limits_{r,s}\left[\frac{\partial\lambda^{(1)}}{\partial r} \exp\{g_1(\hat{r}_o, \hat{s}_o) - g_1(\hat{r}_o, s)\}\right] \max\left(\frac{\partial r}{\partial x}\right)_{t=0}}, \tag{3.92}$$

and

$$\frac{-1}{\max\limits_{r,s}\left[\frac{\partial\lambda^{(2)}}{\partial s} \exp\{g_2(\hat{r}_o, \hat{s}_o) - g_2(r, \hat{s}_o)\}\right] \max\left(\frac{\partial s}{\partial x}\right)_{t=0}}. \tag{3.93}$$

and

t_{sup} is the least positive number of the two quantities

$$\frac{-1}{\min\limits_{r,s}\left[\dfrac{\partial\lambda^{(1)}}{\partial r}\exp\{g_1(\hat{r}_o,\hat{s}_o)-g_1(\hat{r}_o,s)\}\right]\max\left(\dfrac{\partial r}{\partial x}\right)_{t=0}}, \tag{3.94}$$

and

$$\frac{-1}{\min\limits_{r,s}\left[\dfrac{\partial\lambda^{(2)}}{\partial s}\exp\{g_2(\hat{r}_o,\hat{s}_o)-g_2(r,\hat{s}_o)\}\right]\max\left(\dfrac{\partial s}{\partial x}\right)_{t=0}}. \tag{3.95}$$

A further simplification is possible, without actually replacing r and s by the constant values $\hat{r}_o$, $\hat{s}_o$, if the functions $g_1(r,s)$ and $g_2(r,s)$ are approximated by the first two terms of their Taylor series expansions, and the defining relation (3.69) for $g_1(r,s)$ and the corresponding relation for $g_2(r,s)$ are used. To see this write

$$g_1(\hat{r}_o,s) = g_1(\hat{r}_o,\hat{s}_o) + (s-\hat{s}_o)\left(\frac{\partial g_1}{\partial s}\right)_o + O((s-\hat{s}_o)^2)$$

which, by virtue of equation (3.69), becomes

$$g_1(\hat{r}_o,\hat{s}_o) - g_1(\hat{r}_o,s) = \left(\frac{\hat{s}_o - s}{\lambda_o^{(1)}-\lambda_o^{(2)}}\right)\left(\frac{\partial\lambda^{(1)}}{\partial s}\right)_o + O((s-\hat{s}_o)^2),$$

where the suffix o refers to the initial values. A similar result may be obtained for $g_2(r,s)$. Consequently, when r and s differ only slightly from the constant values $\hat{r}_o$ and $\hat{s}_o$ the estimates of t_{inf} and t_{sup} reduce to:

t_{inf} is the least positive number of the two quantities

$$\frac{-1}{\max\limits_{r,s}\left[\dfrac{\partial\lambda^{(1)}}{\partial r}\exp\left\{\left(\dfrac{\hat{s}_o - s}{\lambda_o^{(1)}-\lambda_o^{(2)}}\right)\left(\dfrac{\partial\lambda^{(1)}}{\partial s}\right)_o\right\}\right]\cdot\max\left(\dfrac{\partial r}{\partial x}\right)_{t=0}}, \tag{3.96}$$

and

$$\frac{-1}{\max\limits_{r,s}\left[\dfrac{\partial\lambda^{(2)}}{\partial s}\exp\left\{\left(\dfrac{\hat{r}_o - r}{\lambda_o^{(2)}-\lambda_o^{(1)}}\right)\left(\dfrac{\partial\lambda^{(2)}}{\partial r}\right)_o\right\}\right]\cdot\max\left(\dfrac{\partial s}{\partial x}\right)_{t=0}}, \tag{3.97}$$

and

t_{sup} is the least positive number of the two quantities

$$\frac{-1}{\min_{r,s}\left[\frac{\partial\lambda^{(1)}}{\partial r}\exp\left\{\left(\frac{\hat{s}_o - s}{\lambda_o^{(1)} - \lambda_o^{(2)}}\right)\left(\frac{\partial\lambda^{(1)}}{\partial s}\right)_o\right\}\right]\max\left(\frac{\partial r}{\partial x}\right)_{t=0}}, \tag{3.98}$$

and

$$\frac{-1}{\min_{r,s}\left[\frac{\partial\lambda^{(2)}}{\partial s}\exp\left\{\left(\frac{\hat{r}_o - r}{\lambda_o^{(2)} - \lambda_o^{(1)}}\right)\left(\frac{\partial\lambda^{(2)}}{\partial r}\right)_o\right\}\right]\cdot\max\left(\frac{\partial s}{\partial x}\right)_{t=0}}. \tag{3.99}$$

It is clear from these expressions that the solution will only break down due to the $C^{(1)}$ characteristics when $\max(\partial r/\partial x)_{t=0} > 0$ and, similarly, it will only break down due to the $C^{(2)}$ characteristics when $\max(\partial s/\partial x)_{t=0} > 0$.

A simplified version of these results was obtained by Lax [65] using different comparison theorems, in the last stages of which he assumed that $r = \hat{r}_o$, $s = \hat{s}_o$ in order to study the existence and non-existence of the solutions of a certain nonlinear string equation. Related to this same problem are the pioneering paper by Nitsche [68] and the ingenious approach developed by Ames [69] that is useful for a certain class of problems.

In conclusion it is appropriate to comment on the significance to this analysis of the genuinely nonlinear condition. As all elements of $(\nabla_u\lambda^{(1)})$ and $(\nabla_u\lambda^{(2)})$ appear in estimates (3.96) to (3.99) it will suffice to consider in detail only the partial derivative $\partial\lambda^{(1)}/\partial s$ in (3.96). This partial derivative implies r = const, or β = const, so that it is in fact a directional derivative of $\lambda^{(1)}$ along the $C^{(1)}$ characteristics. Similarly, $\partial\lambda^{(2)}/\partial r$ is a directional derivative of $\lambda^{(2)}$ along the $C^{(2)}$ characteristics. Consequently the vanishing of $\partial\lambda^{(1)}/\partial s$ implies exceptional behaviour with respect to the $C^{(1)}$ characteristic field, and the vanishing of $\partial\lambda^{(2)}/\partial r$ implies exceptional behaviour with respect to the $C^{(2)}$ characteristic field. Thus the genuinely nonlinear condition enters automatically when we assume either, or both, of these derivatives to be non-zero.

3.6 Gas Motion in a Closed Tube

By way of illustration of the results of the previous section consider their application to Example 3 on page 76 which, it will be recalled, was a problem originally due to Riemann [44]. This involved determining the length of time for which smooth one-dimensional motion can exist in a polytropic gas contained between the two fixed walls $x = 0$ and $x = \ell$, when the initial density and velocity variation are specified. As this problem is a mixed initial and boundary value problem it will require conversion to a pure initial value problem before the results of Section 3.5 can be used. This will be accomplished later by a suitable periodic extension of the solution from the fundamental interval $[0,\ell]$ to the entire initial line.

The system of equations involved has already been given in (2.67a,b), while the polytropic gas law connecting pressure and density is

$$p = A\rho^{\gamma}, \tag{3.100}$$

with A and γ positive constants. The Riemann invariants for this system have already been found in equations (3.25a,b) on page 88 without recourse to integrating factors (i.e. by setting $\mu_1 = \mu_2 = 1$). However, as we saw in conjunction with (3.84) that $\partial\lambda^{(1)}/\partial r$ and $\partial\lambda^{(2)}/\partial s$ should be taken to be negative we set $\mu_1 = -1$ and $\mu_2 = 1$ so that equations (3.25a,b) become, respectively,

$$-r = \frac{2a}{\gamma-1} + u \quad \text{and} \quad s = \frac{2a}{\gamma-1} - u, \tag{3.101}$$

giving

$$r + s = -2u \quad \text{and} \quad r - s = \frac{-4a}{(\gamma-1)}. \tag{3.102}$$

Differentiating (r + s) partially with respect to r then gives

$$\frac{\partial u}{\partial r} = -\tfrac{1}{2}, \tag{3.103}$$

while differentiating (r - s) partially with respect to r gives

$$1 = \frac{-4}{\gamma-1}\left(\frac{da}{d\rho}\right)\left(\frac{\partial\rho}{\partial r}\right). \tag{3.104}$$

So as $da/d\rho = (a/2\rho)(\gamma-1)$, and $\lambda^{(1)} = u + a$, we have

$$\frac{\partial\lambda^{(1)}}{\partial r} = \frac{\partial u}{\partial r} + \frac{da}{d\rho}\frac{\partial\rho}{\partial r}, \tag{3.105}$$

from which it follows by combining equations (3.103) to (3.105) that

$$\frac{\partial\lambda^{(1)}}{\partial r} = -\frac{\gamma+1}{4}. \tag{3.106}$$

As it happens this proves to be a constant and so is independent of both r and s. A similar argument shows

$$\frac{\partial\lambda^{(1)}}{\partial s} = \frac{\gamma-3}{4}, \tag{3.107}$$

which is also a constant and independent of r and s.

It is at this point in the argument that the problem requires conversion to an equivalent initial value problem. Preparatory to accomplishing this by a simple periodicity argument we first take note of the fact that the specification of the initial values of u and ρ on $[0,\ell]$ determines the initial values of $r_o(x) = r(x,0)$ and $s_o(x) = s(x,0)$ on this interval. The boundary conditions that are to be imposed on $u(x,t)$ are $u(0,t) = u(\ell,t) = 0$ for all t. So, by virtue of the result $r + s = -2u$, we may write

$$r_o(0) + s_o(0) = r_o(\ell) + s_o(\ell) = 0.$$

Accordingly we extend $r_o(x) + s_o(x)$ to the interval $[-\ell,\ell]$, and thence to the entire initial line, by defining it to be an even function in $[-\ell,\ell]$ that is periodic with period 2ℓ. The function $r_o(x) - s_o(x)$ is similarly extended as an odd function in $[-\ell,\ell]$ which is periodic with period 2ℓ. The boundary conditions on u may now be disregarded since with this extension of $r_o(x)$ and $s_o(x)$ the problem becomes a pure initial value problem.

Considering the pure initial value problem just defined and using the estimates (3.96) and (3.97) shows that when $\max\,(\partial r/\partial x)_{t=0}$ and $\max\,(\partial s/\partial x)_{t=0}$ are positive, t_{inf} is the lesser of the two numbers

$$\frac{4}{(\gamma+1)\max_s\left[\exp\left\{\frac{(\gamma-3)(\hat{s}_o - s)}{8a_o}\right\}\right]\cdot\max\left[\frac{\partial r_o(x)}{\partial x}\right]}, \tag{3.108}$$

and

$$\frac{4}{(\gamma+1)\max_r\left[\exp\left\{\frac{(3-\gamma)(\hat{r}_o - r)}{8a_o}\right\}\right]\cdot\max\left[\frac{\partial s_o(x)}{\partial x}\right]}. \tag{3.109}$$

Similarly from estimates (3.98) and (3.99) we find that t_{sup} is the lesser of the two numbers

$$\frac{4}{(\gamma+1)\min_s\left[\exp\left\{\frac{(\gamma-3)(\hat{s}_o - s)}{8a_o}\right\}\right]\cdot\max\left[\frac{\partial r_o(x)}{\partial x}\right]}, \tag{3.110}$$

and

$$\frac{4}{(\gamma + 1) \min_r \left[\exp\left\{ \frac{(3 - \gamma)(\hat{r}_o - r)}{8a_o} \right\} \right] \cdot \max \left[\frac{\partial s_o(x)}{\partial x} \right]} . \tag{3.111}$$

When r and s in these expressions are replaced by their associated constan values $\hat{r}_o$ and $\hat{s}_o$, as was done by Ludford [45], so that $r = \hat{r}_o$, $s = \hat{s}_o$, the numbers t_{inf} and t_{sup} coincide. Denoting by β the greater of the two quantities $\max[\partial r_o(x)/\partial x]$ and $\max[\partial s_o(x)/\partial x]$ then yields the simplest asymptotic estimate

$$t_c = \frac{4}{(\gamma + 1)\beta}$$

for the time of breakdown of the solution.

This is precisely the result obtained by Ludford when allowance is made for the fact that his definitions of the Riemann invariants r and s differ by a numerical factor two from those given in equations (3.101).

Further applications of Section 3.5 to other problems are to be found in reference [46]. These include the breaking of waves in shallow water and the propagation of electromagnetic waves and the formation of shocks in a nonlinear electromagnetic transmission line. When the estimates (3.96) to (3.99) are applied to the nonlinear string problem studied by Lax [65] which has motion governed by an equation of the form (1.8), with y replaced by t, his result is recovered when, as did Lax, we set $r = \hat{r}_o$ and $s = \hat{s}_o$.

3.7 Unboundedness of Solutions

Sections 3.5 and 3.6 have been concerned with the breakdown of the differentiability of solutions to reducible systems. The purpose of this last section is to show that when systems involving more than three equations are involved it is also possible for the solution itself to become infinite a finite elapsed time after the start. This is achieved by way of an example first constructed by Jeffrey in reference [11], which although quasilinear and hyperbolic also happens to be completely exception in the sense of Definition 3.4

In addition to the unbounded nature of the solution after a finite elapsed time, rather than the unboundedness of its derivative, this example is of interest because it involves a family of solutions U_μ which depend continuousl on a non-negative parameter μ. In the limit as $\mu \to 0$ so the system and its solution vector U_μ approach uniformly a linear hyperbolic system and its solution vector U_o which has the same characteristic fields. The system of equations involved has already been encountered in (1.19a,b), while its hyperbolicity was established in sub-section 6 on page 56 and its completey exceptional nature at the end of Section 3.4. The system was, in fact, suggested by a special example, not involving a family of solutions, that has been attributed to V.I. Zurkov by Rozhdestvensky [70]. As the system cannot be expressed in divergence form it is not a conservative system as defined in equation (1.11).

The Initial Value Problem

Let $u_{1\mu}$, $u_{2\mu}$, and $u_{3\mu}$ be the components of a 3×1 vector U_μ depending on a real non-negative scalar parameter μ and satisfying the equation

$$\frac{\partial U_\mu}{\partial t} + A(\mu,U_\mu)\frac{\partial U_\mu}{\partial x} = 0, \tag{3.112a}$$

where

$$A(\mu,U_\mu) = \begin{bmatrix} -\cosh 2\mu u_{2\mu} & 0 & -\sinh 2\mu u_{2\mu} \\ \cosh \mu u_{2\mu} & 0 & \sinh \mu u_{2\mu} \\ \sinh 2\mu u_{2\mu} & 0 & \cosh 2\mu u_{2\mu} \end{bmatrix}. \tag{3.112b}$$

The initial value problem we now propose for system (3.112) requires that on an interval I of the initial line the elements of U_μ satisfy the Cauchy data

$$u_{1\mu}(x,0) = x/\alpha h, \quad u_{2\mu}(x,0) = 0, \quad u_{3\mu}(x,0) = -x/\alpha h, \tag{3.113}$$

with $\alpha > 0$ and I defined as the interval $x \in [-h,h]$.

The three eigenvalues of $A(\mu,U_\mu)$ determined by

$$|A(\mu,U_\mu) - \lambda I| = 0, \tag{3.114}$$

have already been seen to be $\lambda^{(1)} = -1, \lambda^{(2)} = 0$, and $\lambda^{(3)} = 1$. They are thus independent of both the elements of the solution vector U_μ and the parameter μ itself. The field of characteristic curves $C^{(i)}$ corresponding to the eigenvalue $\lambda^{(i)}$ is determined by the family of solutions to the different equation

$$C^{(i)} : dx/dt = \lambda^{(i)}, \quad \text{for } i = 1,2,3. \tag{3.115}$$

Each field of characteristic curves thus comprises a family of straight lines with gradients -1,0, and 1, respectively.

Domain of Determinacy and Unboundedness of Solution

It is easily established that the vector U_μ with elements $u_{1\mu}$, $u_{2\mu}$ and $u_{3\mu}$ defined as follows satisfies system (3.112a) and coincides with the Cauchy data (3.113) on the initial line:

$$u_{1\mu}(x,t) = \frac{1}{\mu}\left[\frac{\alpha h}{\alpha h - \alpha t} + \frac{\mu x}{\alpha h} - 1\right], \tag{3.116a}$$

$$u_{2\mu}(x,t) = \frac{1}{\mu}\left[\log\left|1 - \frac{\mu t}{\alpha h}\right|\right], \tag{3.116b}$$

$$u_{3\mu}(x,t) = \frac{1}{\mu}\left[\frac{\alpha h}{\alpha h - \mu t} - \frac{\mu x}{\alpha h} - 1\right]. \tag{3.116c}$$

This vector U_μ will only be the solution vector for this quasilinear hyperbolic system when x,t lie within the domain of determinacy of the solution, that is, the region in the (x,t)-plane lying above the initial line and bounded by the domain of dependence I and the two extreme backward drawn characteristics from a point in the upper half plane that pass through the end points of I. As the families of characteristic curves for system (3.112a) are straight lines, these extreme characteristics are the lines with gradients 1 and -1 passing, respectively, through the points (-h,0) and (h,0). The domain of determinacy is shown as the shaded region $\mathcal{G}$ in Figure 17. Whenever U_μ is defined for some $t = t_1$, where $t_1 \leq h$, the Cauchy data will determine, through U_μ, the solution at every point of the line AB, since this lies within the domain of determinacy $\mathcal{G}$ of the solution. The vector U_μ will also determine the solution at all points in $\mathcal{G}$ below the line AB. However, when $t = t_2$ with $t_2 > h$, the corresponding line between the points C and D in Figure 17 has no intersection with $\mathcal{G}$. Under these circumstances the vector U_μ cannot represent the solution along any such line.

The exceptional behaviour of this quasilinear hyperbolic system of three homogeneous equations is seen to be due to the fact that the three eigenvalues of the matrix $A(\mu, U_\mu)$ are constant. However, as this result also implies that each family of characteristic curves is a family of parallel straight lines this also means that no breakdown of the solution can occur due to the intersection of characteristics belonging to any one family, as happened in the example considered on page 32. This is an important conclusion to which we shall have occasion to refer to again on page 126. Let us now study the nature of solution (3.116) to our initial value problem.

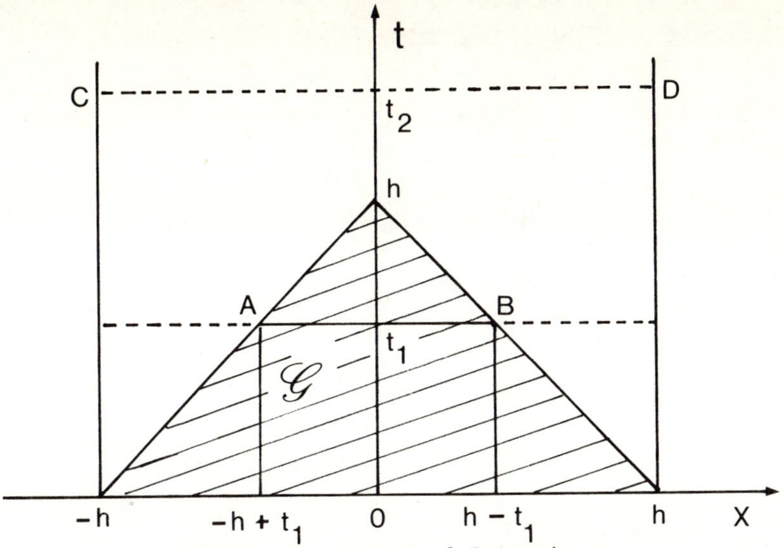

Figure 17: Domain of Determinacy

An examination of the form of results (3.116) shows that U_μ becomes unbounded when $\alpha h = \mu t$. This defines a critical time t_c through the relation

$$t_c = \alpha h/\mu, \tag{3.117}$$

with the property that the vector U_μ will only be a solution vector, and also finite, for points $(x,t) \in \mathscr{G}$ for which $t < t_c$.

Consider first the case in which $\mu < \alpha$. Then it follows from (3.117) that $t_c > h$, so that we have a situation corresponding to the unboundedness of the vector U_μ occurring along a line, similar to CD, which does not intersect $\mathscr{G}$. Thus, when $\mu < \alpha$, the vector U_μ with elements (3.116) represents the solution at every point of $\mathscr{G}$, and at no point of $\mathscr{G}$ does the solution become unbounded.

Next, consider the case in which $\alpha \le \mu$. It then follows from (3.117) that $t_c \le h$, so that we have a situation corresponding to the unboundedness of the vector U_μ occurring along a line similar to AB which lies in $\mathscr{G}$. Under these circumstances U_μ will only represent the solution vector for those points $(x,t) \in \mathscr{G}$ which also lie below AB. In summary the situation is as follows:

(i) The case $\mu < \alpha$. The domain of determinacy $\mathscr{G}$ comprises the set of points

$$\mathscr{G} = \{x,t \,|\, |x| \le h, t \ge 0, t \le h - x, t \le h + x\}.$$

The vector $U_\mu(x,t)$ has the elements defined in (3.116), and represents the solution vector at every point of $\mathscr{G}$, at no point of which does it become unbounded.

(ii) The Case $\alpha \le \mu$. The domain of determinacy $\mathscr{G}$ comprises the set of points

$$\mathscr{G} = \{x,t \,|\, |x| \le h, t \ge 0, t \le h - x, t \le h + x\}.$$

The vector $U_\mu(x,t)$ has the elements defined in (3.116), but only represents the solution vector at the subset $\mathscr{G}_1$ of $\mathscr{G}$, defined by the relation

$$\mathscr{G} = \{x,t \,|\, |x| \le h, t \ge 0, t \le h - x, t \le h + x, t < \alpha h/\mu\}.$$

Along the line segment L, comprising the set of points

$$L = \{x,t \,|\, |x| \le h(1 - \alpha/\mu), t = \alpha h/\mu\},$$

the solution vector U_μ becomes unbounded.

The unbounded behaviour of the solution vector is in no way attributable to the intersection of the characteristic curves of any one of the three families involved, since within each family $C^{(i)}$, $i = 1,2,3$, the characteristic curves are parallel straight lines.

Limiting Case as $\mu \to 0$

We now examine the behaviour of system (3.112) and its solution vector U_μ as $\mu \to 0$. Proceeding to the limit as $\mu \to 0$ in expressions (3.116) gives:

$$u_{10}(x,t) = (t + x)/\alpha h, \tag{3.118a}$$

$$u_{20}(x,t) = -t/\alpha h, \tag{3.118b}$$

$$u_{30}(x,t) = (t - x)/\alpha h. \tag{3.118c}$$

Differentiation establishes that the vector $U_o(x,t)$ with elements u_{10}, u_{20} and u_{30} satisfies the linear hyperbolic system

$$\frac{\partial U_o}{\partial t} + A_o \frac{\partial U_o}{\partial x} = 0, \tag{3.119a}$$

where

$$A_o = \begin{bmatrix} -1 & 0 & 0 \\ 1 & 0 & 0 \\ 0 & 0 & 1 \end{bmatrix}. \tag{3.119b}$$

Since the eigenvalues of A_o are the same as those of $A(\mu, U_\mu)$, the characteristic curves of system (3.119) will be the same as those of system (3.112). It is also true that $U_o(x,0)$ satisfies the Cauchy data (3.113) so that the domains of determinacy of both the linear and the quasilinear systems will be the same domain $\mathscr{G}$ defined above. Furthermore, from the linear theory we know that U_o is the solution vector of (3.119) throughout $\mathscr{G}$, and at each point of $\mathscr{G}$ vector U_o is finite. These results, taken together with the fact that $A_o = \lim_{\mu\to 0} A(\mu, U_\mu)$, establish that the solution vector U_o to the linear hyperbolic system (3.119) subject to the Cauchy data (3.113) is the uniformly continuous limit as $\mu \to 0$ of the solution vector U_μ to the quasilinear hyperbolic system (3.112), subject to the same initial data. This result is surprising and is, perhaps, as interesting as the unbounded behaviour of U_μ itself.

To examine the relationship of this result to the conclusions reached above it is only necessary to consider Case (i). This is because, as α is a fixed positive constant, when proceeding to the limit as $\mu \to 0$ we will eventually satisfy the condition $\mu < \alpha$. When this occurs U_μ will be defined, and finite, throughout all of $\mathscr{G}$, as is U_c, to which it tends uniformly as $\mu \to 0$. The elements of U_o, will only become unbounded at infinity, though, since I is finite, U_o will not then represent the solution vector.

The general question of the existence and non-existence of solutions to partial differential equations has already given rise to a considerable amount of literature, part of which is related to improperly posed problems. A number of contributions to this general area have been provided in the collection of papers edited by Knops [82], in which particular attention should be drawn to the papers by L.E. Payne, R.J. Knops and H.A. Levine, and also to their bibliographies. For more specific references to non-existence theorems of a very general kind that may be proved and for further references we mention here only two papers by these same authors [83,84].

4 Shock waves

4.1 Conservation Systems and Conditions Across a Shock

In what follows it will be assumed that the system of equations involved is hyperbolic and capable of expression in the generalised conservation form first introduced in (1.11). That is, when the system involves n dependent variables and is formulated in $\mathbb{R}^3 \times t$, we assume it can be written in the divergence form

$$\frac{\partial F}{\partial t} + \text{div } G = H, \tag{4.1}$$

with $U = U(\underline{x},t)$, $F = F(U(\underline{x},t))$ and $H = H(U(\underline{x},t))$ all n element column matrix vectors and $G = G(U(\underline{x},t))$ an $n \times 3$ matrix. The matrix G in (4.1) is in effect to be regarded as a tensor so that div G has the meaning

$$\text{div } G = \sum_{s=1}^{3} \frac{\partial g^{(s)}}{\partial x_s}, \tag{4.2}$$

where $g^{(s)}$ is the s-th column of G.

Systems of this type are of considerable importance because of their frequent occurrence in physical problems where they arise from integral formulations of quantities that are conserved. Indeed, since an integral formulation is more fundamental than the related differential equation and it permits the integrand to be discontinuous, we shall make use of it to discuss discontinuous solutions for system (4.1).

Discontinuous solutions have considerable physical significance, since they may be interpreted in terms of physical phenomena such as a shock wave in a gas and a bore in shallow water, to mention but two possible examples. If a discontinuous solution exists across a surface, the first problem to be resolved is how the solutions on adjacent sides of the surface are to be related one to the other and to the speed of propagation of the surface. In the case of a shock wave in a gas this involves determining the relationship connecting gas pressures and densities on opposite sides of the shock with the speed of propagation of the shock. Of the various ways in which this may be achieved we choose to base our approach on that given in the paper by Jeffrey [71] which uses the following simple theorem from vector field theory.

Theorem 4.1 (Integral Rate of Change Theorem)

Let F be an $n \times 1$ column matrix with elements which are continuous scalar functions of position and time defined throughout the volume V(t), which is itself bounded by a surface S(t) moving with velocity $\underline{v}$. Then the rate of change of the volume integral of F is given by

$$\frac{d}{dt}\int_{V(t)} F\,dV = \int_{V(t)} \frac{\partial F}{\partial t}\,dV + \int_{S(t)} F\,\underline{v}.d\underline{S},$$

where $d\underline{S}$ is the vector element of surface area.

Proof

Let f_i be an element of F, and set

$$I = \int_{V(t)} f_i\,dV. \tag{4.3}$$

Then in time increment δt the integrand of (4.3) becomes, to the first order,

$$f_i + \left\{\frac{\partial f_i}{\partial t}\right\}\delta t. \tag{4.4}$$

However, during this time increment δt, the volume V(t) bounded by S(t) changes due to the motion of S(t) itself. The effect of this on the volume integral of F follows by observing that if the outward drawn vector surface element of S(t) is $d\underline{S}$, then in time increment δt the element moves a distance $\underline{v}\delta t$, with a corresponding volume element of change $\underline{v}.d\underline{S}\,\delta t$. The corresponding increment in the integrand of (4.3) due to this is thus $f_i\underline{v}.d\underline{S}\,\delta t$. Consequently, combining these results we find that the increment δI in I is given by

$$I + \delta I = \int_{V(t+\delta t)} [f_i + \left\{\frac{\partial f_i}{\partial t}\right\}\delta t]dV + \int_{S(t+\delta t)} f_i\,\underline{v}.d\underline{S}\,\delta t. \tag{4.5}$$

Finally, subtracting (4.3) from (4.5), dividing by δt and proceeding to the limit as $\delta t \to 0$ we obtain the result

$$\frac{d}{dt}\int_{V(t)} f_i \, dV = \int_{V(t)} \frac{\partial f_i}{\partial t} \, dV + \int_{S(t)} f_i \, \underline{v}.d\underline{S}. \tag{4.6}$$

The statement of the theorem then follows by application of (4.6) to each element f_i of F. ∎

To apply this theorem to discontinuous solutions of system (4.1) it is first necessary to express the system in integral form, and it is at this point that the divergence term becomes significant, because appeal may be made to the Gaussian divergence theorem to convert a volume integral to a surface integral. If volume V(t) is then chosen so that it is divided into two parts by the discontinuity surface, the required discontinuity condition follows by subtracting from the integral over V(t) the integrals over the two sub-volumes, thereby leaving only the difference of two surface integrals across the discontinuity.

Let us now identify the column matrix F in Theorem 4.1 with the $n \times 1$ column matrix F in system (4.1) and assume that a surface $\sigma(\underline{x},t) = \text{const}$ exists across which the matrix vector U, and hence F, G and H are discontinuous. Next we choose the volume V(t) bounded by surface S(t) moving with velocity $\underline{v}$ so that an arbitrary part $S_o(t)$ of the discontinuity surface $\sigma(\underline{x},t) = \text{const}$ divides it into the two sub-volumes $V_+(t)$ and $V_-(t)$. Denote by $S_+(t)$ and $S_-(t)$ those parts of S(t) that bound $V_+(t)$ and $V_-(t)$, respectivley, excluding the dividing surface $S_o(t)$ which, we assume, also has velocity $\underline{v}$

Integrating (4.1) over $V(t) = V_+(t) \cup V_-(t)$ gives

$$\int_{V_+ \cup V_-} \frac{\partial F}{\partial t} \, dV + \int_{V_+ \cup V_-} \text{div } G \, dV = \int_{V_+ \cup V_-} H \, dV$$

or, from the matrix form of the Gaussian divergence theorem applied separately to V_+ and V_- in which F, G are continuous and differentiable,

$$\int_{V_+ \cup V_-} \frac{\partial F}{\partial t} \, dV + \int_{S_+ \cup S_-} G.d\underline{S} = \int_{V_+ \cup V_-} H \, dV, \tag{4.7}$$

where $G.d\underline{S}$ denotes the scalar product of G now regarded as a tensor and vector $d\underline{S}$. Combining (4.7) with the result of Theorem 4.1 applied separately to V_+ and V_- then gives the next result in which, it must be remembered, the dividing surface $S_o(t)$ that is part of $\sigma(\underline{x},t) = \text{const}$ also moves with velocity $\underline{v}$

$$\frac{d}{dt}\int_{V_+\cup V_-} F dV = \int_{S_+\cup S_-} (F\underline{v} - G).d\underline{S} + \int_{V_+\cup V_-} H dV. \tag{4.8}$$

If, now, we subtract from (4.8) the corresponding expressions integrated over the separate volumes $V_+(t)$ and $V_-(t)$, and bounded, respectively, by $S_+(t)\cup S_o(t)$ and $S_-(t)\cup S_o(t)$ we arrive at the result

$$\int_{S_o(t)} (F\underline{v} - G)_+ . d\underline{S}_+ + \int_{S_o(t)} (F\underline{v} - G)_- . d\underline{S}_- = 0, \tag{4.9}$$

where $d\underline{S}_+$ and $d\underline{S}_-$ are the outward directed surface elements with respect to the volumes $V_+(t)$ and $V_-(t)$. This situation is illustrated diagramatically in Fig. 18 which shows an arbitrarily thin volume element taken across $\sigma(\underline{x},t)$ = const. The effect of differencing to obtain (4.9) is to make the volume contribution and the contribution due to the surface element $d\underline{S}'$ directed along $\underline{n}'$ parallel to $\sigma(\underline{x},t)$ = const vanish in the limit as the cylinder collapses onto the area element dS_o.

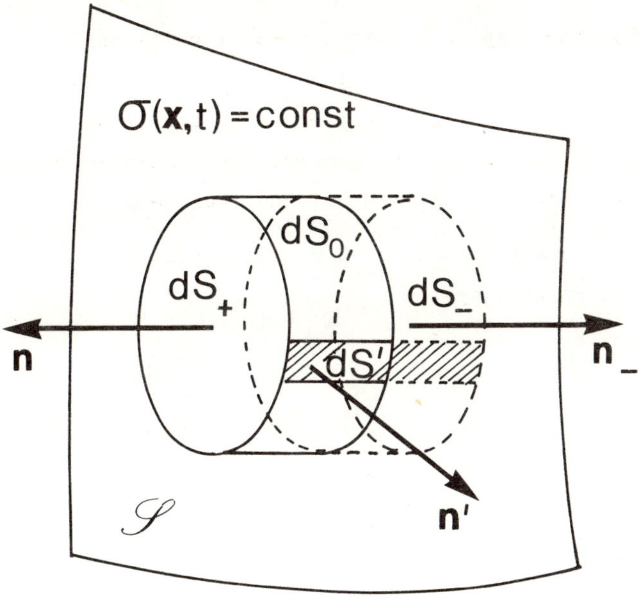

Fig. 18. Volume element divided by discontinuity surface $\sigma(\underline{x},t)$ = const.

Since $d\underline{S}_\pm$ are both normal to the same discontinuity surface $\sigma(\underline{x},t)$ = const, but are oppositely directed so that $\underline{n} = -\underline{n}_-$, we have

$d\underline{S}_+ = -d\underline{S}_- = \underline{n}dS_o$ showing that (4.9) may be re-written as

$$\int_{S_o(t)} \{(F\underline{v} - G)_+.\underline{n} - (F\underline{v} - G)_-.\underline{n}\}dS_o = 0. \tag{4.10}$$

The fact that dS_o is arbitrary then gives an algebraic jump condition across $\sigma(\underline{x},t) = \text{const}$ of the form

$$(F\underline{v} - G)_+.\underline{n} - (F\underline{v} - G)_-.\underline{n} = 0. \tag{4.11}$$

It is useful to re-express this result by observing that the scalar quantities $\underline{v}_+ .\underline{n}$ and $\underline{v}_- .\underline{n}$ are the normal speeds of propagation of the elements $d\underline{S}_+$ and $d\underline{S}_-$ on opposite sides of, and moving with, $\sigma(\underline{x},t) = \text{const}$, and as such must be continuous across $S_o(t)$. So writing $\tilde{\lambda} = \underline{v}_+.\underline{n} = \underline{v}_- .\underline{n}$ enables the jump condition (4.11) to be expressed in an alternative form using the speed $\tilde{\lambda}$ normal to $\mathscr{S}$

$$\tilde{\lambda}(F_+ - F_-) = (G_+ - G_-).\underline{n}, \tag{4.12}$$

which is sometimes written

$$\tilde{\lambda}[F] = [G].\underline{n}, \tag{4.13}$$

with $[Q]$ denoting the jump in Q across discontinuity surface $S_o(t)$. The arbitrary nature of dS_o also implies that $U_\pm$ varies continuously over $\mathscr{S}$. Because of the similarity of (4.12) to a corresponding condition in gas dynamics [9] this result will be called the generalised Rankine-Hugoniot condition for system (4.1). The following result has thus been proved.

Theorem 4.2 (Generalised Rankine-Hugoniot Condition)
Consider the conservation system

$$\frac{\partial F}{\partial t} + \text{div } G = H,$$

with $F = F(U(\underline{x},t))$, $G = G(U(\underline{x},t))$ and $H = H(U(\underline{x},t))$. Then, if this has a discontinuous solution across a surface $\mathscr{S}$, on the adjacent sides $\pm$ of $\mathscr{S}$ the solution varies continuously and is related by the jump condition

$$\tilde{\lambda}(F_+ - F_-) = (G_+ - G_-).\underline{n},$$

in which $\underline{n}$ is the normal to $\mathscr{S}$ and $\tilde{\lambda}$ is the normal speed of propagation of $\mathscr{S}$, with $G_\pm$ regarded as a tensor and $G_\pm.\underline{n}$ denoting the scalar product of G and the unit vector $\underline{n}$ normal to $\mathscr{S}_\pm$. ▌

Again by analogy with gas dynamics, a discontinuous solution to a hyperbolic conservation system will be called a"shock". Although discontinuous solutions also exist for hyperbolic systems which are non-conservative, the manner in which the solutions on either side of the discontinuity surface are to be related is not then clear. Accordingly, and to avoid confusion, we will simply refer to a discontinuous solution to an arbitrary hyperbolic system as a "strong discontinuity"

Definition 4.1 (Shock Solution)
A discontinuous solution to a system of equations expressed in conservation form which satisfies the generalised Rankine-Hugoniot condition will be called a shock. ▮

The jump condition (4.12) is a pointwise condition across the shock surface $\mathcal{S}$ and, since F, G and H depend on the n element column vector U, it comprises n nonlinear algebraic equations. If the solution vectors on either side of $\mathcal{S}$ are denoted by U_+ and U_-, we see that (4.12) provides n nonlinear equations for the 2n + 1 quantities comprising the 2n elements of U_+ and U_- and the shock speed $\tilde{\lambda}$.

Often, by analogy with gas dynamics, the solutions $U_\pm$ are termed the "solution states" on either side of $\mathcal{S}$, since in physical problems the elements of U are the state variables occurring in system (4.1).

The continuation of a differentiable solution across a shock $\mathcal{S}$ by using the generalised Rankine-Hugoniot condition (4.12) is not necessarily a unique process. This may occur because of the nonlinearity of F and G and also because the jumps in the elements of U may be either positive or negative. Any discontinuous solution satisfying (4.12) will be a mathematical solution to (4.1), though if system (4.1) is derived from a physical problem the discontinuous solution will not necessarily be physically realisable. We return to this problem later on in this chapter.

Let us now examine the restriction of condition (4.12) to one space variable and time when a considerable simplification results. In this case G becomes a column matrix and $\underline{n}$ merely the unit vector along the x-axis so that (4.12) reduces to

$$\tilde{\lambda}(F_+ - F_-) = (G_+ - G_-), \tag{4.14}$$

where $\tilde{\lambda}$ is now the shock speed along the x-axis. This is the form in which we shall need to use the generalised Rankine-Hugoniot condition in both the general theory and the application in Chapter 5.

It is important to recognise that, in general, only when system (4.1) is linear or semi-linear will the shock speed $\tilde{\lambda}$ be the same as the normal speed of propagation of a characteristic surface. We show this only in the case of one space dimension and time, though it is readily proved for more space dimensions. It will suffice to consider a hyperbolic semi-linear system which may be written

$$\frac{\partial U}{\partial t} + \frac{\partial G}{\partial x} = H \tag{4.15}$$

with $G = AU$, where A is independent of U, but $H = H(U,x,t)$. Then making the identifications $F \to U$, $G \to AU$ in (4.14) and using the fact that the independence of A from U implies A is continuous across the discontinuity surface leads to the result

$$\tilde{\lambda}(U_+ - U_-) = A(U_+ - U_-),$$

or to

$$(A - \tilde{\lambda}I)(U_+ - U_-) = 0.$$

If a shock exists, so that $U_+ - U_- \neq 0$, then this is only possible if

$$|A - \lambda I| = 0,$$

which is merely the characteristic determinant for (4.15) which may also be written as

$$\frac{\partial U}{\partial t} + A\frac{\partial U}{\partial x} = H.$$

Consequently in the semi-linear case (and also the linear case) the shock speeds $\tilde{\lambda}$ coincide with the characteristic speeds. Only in exceptional circumstances may this coincidence of characteristics and shocks occur with quasilinear hyperbolic systems [72].

As already mentioned in connection with equation (1.12), when expressed in terms of the matrix vector U, system (4.1) has the form

$$(\nabla_u F)\frac{\partial U}{\partial t} + \sum_{s=1}^{3} (\nabla_u g^{(s)})\frac{\partial U}{\partial x_s} = H,$$

showing that the matrix A associated with (4.15) is just $A = (\nabla_u G)$.

If a general system of equations

$$\frac{\partial U}{\partial t} + \sum_{i=1}^{3} A_i(U,\underline{x},t)\frac{\partial U}{\partial x_i} + B(U,\underline{x},t) = 0$$

is considered, then there is no simple method by which it can be shown to be derived from a system in the conservation form (4.1) and by which F, G and H may be found. Something of the argument necessary in individual cases has been indicated in connection with examples 1, 2 and 4 on pages 10 to 14.

We conclude this section by deriving the connection between the speed of propagation $\tilde{\lambda}$ of the shock $\mathcal{S}$ along its normal and the equation $\sigma(\underline{x},t) = \text{const}$ of the shock surface itself. Taking the total derivative of $\sigma(\underline{x},t) = \text{const}$ leads directly to the equation

$$\frac{\partial\sigma}{\partial t} + (\nabla_x\sigma).\frac{d\underline{\underline{x}}}{dt} = 0,$$

where ∇_x is the spatial gradient operator. Dividing by $|\nabla_x\sigma|$ and setting $\underline{n} = (\nabla_x\sigma)/|\nabla_x\sigma|$, the unit normal to $\mathcal{S}$, we arrive at the result

$$\underline{n}.\frac{d\underline{\underline{x}}}{dt} = -\frac{\partial\sigma}{\partial t}\Big/|\nabla_x\sigma|.$$

Now as $\underline{x}$ is constrained to lie on the shock surface $\mathcal{S}$, the velocity $d\underline{x}/dt$ is the local velocity of propagation of the surface, so that $\underline{n}.(d\underline{x}/dt)$ is the local speed of propagation of $\mathcal{S}$ along its normal. We have thus derived the required connection in the form

$$\tilde{\lambda} = \underline{n}\cdot\frac{d\underline{\underline{x}}}{dt} = -\frac{\partial\sigma}{\partial t}\Big/|\nabla_x\sigma|. \qquad (4.16)$$

4.2 Conservation Equations and Shocks in Fluid Dynamics

Preparatory to examining the way in which solutions to system (4.1) may be extended in a unique manner to admit the class of piecewise differentiable functions separated by shocks it will be useful to examine a specific physical problem. The most convenient mathematical model of a physical problem for this purpose is provided by fluid dynamics [7,8,9,53,85]. Our starting point will be the equations of mass, momentum and energy in an inviscid fluid, which in conservation form may be written:

$$\frac{\partial\rho}{\partial t} + \mathrm{div}(\rho\underline{q}) = 0 \qquad \text{(mass)}, \tag{4.17}$$

$$\frac{\partial(\rho\underline{q})}{\partial t} - \mathrm{div}\,\mathbf{T} = 0 \qquad \text{(momentum)}, \tag{4.18}$$

and

$$\frac{\partial W}{\partial t} + \mathrm{div}\,\mathbf{H} = 0 \qquad \text{(energy)}. \tag{4.19}$$

where the pressure tensor $\mathbf{T}$, the energy flow vector $\mathbf{H}$ and the energy density W are given by

$$T_{ik} = -(p\delta_{ik} + \rho q_i q_k), \tag{4.20}$$

$$\mathbf{H} = \rho\underline{q}(\tfrac{1}{2}\underline{q}^2 + e) + p\underline{q} + \underline{w}, \tag{4.21}$$

and

$$W = \rho(\tfrac{1}{2}\underline{q}^2 + e), \tag{4.22}$$

with ρ the density, $\underline{q}$ the fluid velocity, p the pressure, e the internal energy and $\underline{w}$ the heat flow vector. Applying the scalar form of jump condition (4.12) to (4.17) and identifying ρ with F and $\rho\underline{q}$ with G, we obtain

$$\{\rho(\underline{v}-\underline{q})\}_+.\underline{n} - \{\rho(\underline{v}-\underline{q})\}_-.\underline{n} = 0. \tag{4.23}$$

Similarly, using equation (4.18) and identifying $\rho\underline{q}$ with F and   with G, we find

$$\{\rho\underline{q}(\underline{v}-\underline{q}).\underline{n} - p\underline{n}\}_+ - \{\rho\underline{q}(\underline{v}-\underline{q}).\underline{n} - p\underline{n}\}_- = 0, \tag{4.24}$$

while from (4.19) in which we identify W with F and $-\mathbf{T}$ with G we find

$$\{\rho(\tfrac{1}{2}\underline{q}^2 + e)(\underline{v}-\underline{q})-(p\underline{q}+\underline{w})\}_+.\underline{n} - \{\rho(\tfrac{1}{2}\underline{q}^2 + e)(\underline{v}-\underline{q})-(p\underline{q}+\underline{w})\}.\underline{n} = 0. \tag{4.25}$$

It is now convenient to introduce the symbol Δ to indicate the jump $\Delta h \equiv h_+ - h_-$ in a quantity h, when (4.23) becomes

$$(\underline{v}-\underline{q}_+).\underline{n}(\Delta\rho) = \rho_-(\Delta\underline{q}).\underline{n}. \tag{4.26}$$

Similarly, using (4.23), equations (4.24) and (4.25) may be written, respectively,

$$\rho_+(\underline{v}-\underline{q}_+).\underline{n}(\Delta\underline{q}) = (\Delta p)\underline{n} \tag{4.27}$$

and

$$\rho_+(\underline{v}-\underline{q}_+).\underline{n}\Delta(\tfrac{1}{2}\underline{q}^2 + e) = \Delta(p\underline{q}+\underline{w}).\underline{n}. \quad (4.28)$$

The quantity $(\underline{v}-\underline{q}_+).\underline{n}$ appearing on the left hand side of these equations is just the speed of the moving discontinuity surface relative to the fluid velocity in the + state.

It is at this point that in fluid dynamics the concept of a shock is more restrictive than that allowed by our Definition 4.1. This is because in fluid dynamics a discontinuity is only considered to be a shock if a mass flow takes place across it. Expressed another way, the requirement is that fluid should flow through the discontinuity surface. In terms of results (4.26) to (4.28) this is equivalent to requiring that $(\underline{v}-\underline{q}_+).\underline{n} \neq 0$ and $\Delta\rho \neq 0$ for a shock to occur. For a related approach see reference [73].

Let us see what happens when the discontinuity surface moves with the medium in such a way that $(\underline{v}-\underline{q}_+).\underline{n} = 0$. The above results then become

$$(\Delta\underline{q}).\underline{n} = 0, \quad (4.29)$$

$$\Delta p = 0 \quad (4.30)$$

and

$$(\Delta\underline{w}).\underline{n} = 0. \quad (4.31)$$

This form of discontinuity is usually known in fluid dynamics by the name contact discontinuity. It has the property that the pressure and the normal components of fluid velocity and of the heat flow vector are continuous across it. The physical existence of such a discontinuity ignores the effects of viscosity and so in reality it is unstable and can only exist as an approximation for a limited period of time. Such a discontinuity represents some form of stratified fluid flow. A discussion of this matter under the name Kelvin-Helmholtz instability is to be found in the reference work by Chandrasekhar [74].

It follows from (4.26) and (4.27) that for a shock, which is the stronger form of discontinuity in which $(\underline{v}-\underline{q}_+).\underline{n} \neq 0$ and $\Delta\rho \neq 0$, we have

$$\Delta\underline{q} = \frac{(\Delta p)\underline{n}}{\rho_+(\underline{v}-\underline{q}_+).\underline{n}} \quad (4.32)$$

and

$$(\underline{v}-\underline{q}_+) \cdot \underline{n} = \pm\left(\frac{\rho_+ \Delta p}{\rho_- \Delta \rho}\right)^{\frac{1}{2}} \quad . \qquad (4.33)$$

If, however, $\Delta\rho = 0$ but $(\underline{v}-\underline{q}_+) \cdot \underline{n} \neq 0$ we find from (4.27) and (4.28) that $\underline{q}$ and p are continuous so that

$$(\underline{v}-\underline{q}_+) \cdot \underline{n} = \frac{(\Delta\underline{w}) \cdot \underline{n}}{\rho_+ \Delta e} \quad . \qquad (4.34)$$

This form of discontinuity is called a phase front.

The analysis of the gas shock wave provides an excellent example of the type mentioned on page 133 in which a mathematical solution exists to jump condition (4.12) that is not physically realisable. This comes about because a gas shock wave is known from physical observation to always be compressive, in the sense that gas density increases after the passage of the shock wave. However in the mathematical solution the gas pressure and density may either increase or decrease together. This last result may be seen from (4.33) which, because of the square root, requires Δp and $\Delta\rho$ to both be of the same sign. The difficulty is resolved in this case by appeal to a thermodynamic principle which has not so far been mentioned in connection with equations (4.17) to (4.19). If in addition to satisfying the condition in gas dynamics analogous to (4.12) we assert from the second law of thermodynamics that the entropy must not decrease, then we find that only the compression shock is possible [7,9].

The selection principle used here for identifying a physically realisable gas shock wave is called the entropy condition and it has been imposed from outside the framework of the equations of fluid dyanmics. Without this selection principle the mathematical shock solution would not be unique. This imposition of an external selection principle is a concept to which we shall need to return later when considering the generalised Rankine-Hugoniot condition for general systems (4.1) which have no thermodynamic principle to be invoked in order to resolve such difficulties.

There are numerous other applications of the jump conditions (4.12) to physical problems that are to be found in the literature which also deserve mention. In magnetohydrodynamics accounts are to be found in Jeffrey and Taniuti [8], Bazer and Ericson [31], Kulikovskiy and Lyubimov [75] and Pai [76]; in nonlinear electromagnetic theory we mention Katayev [77]; in general nonlinear elasticity there is the reference work by Eringen and Suhubi [54] and in magnetoelasticity the paper by Bazer and Ericson [78]; while in the study of water waves there are the basic works by Stoker [10, 79]. As general references we finally mention the collection of papers on hyperbolic equations and waves edited by Froissart [80] and the non-mathematical but remarkably illustrated account of shock waves by Glass [81], which makes clear the wide variety of physical phenomena that are connected with the study of shocks.

4.3 Weak Solutions and Non-Uniqueness

So far, in the development of the concept of a solution to a quasilinear hyperbolic system, care has been taken to distinguish between classical once differentiable so called C^1 solutions, and piecewise differentiable C^1 solutions separated by shocks across which both U and its derivatives are discontinuous. It would be desirable, if possible, to unify these two types of solution by generalising the whole concept of a "solution" to system (4.1) in such a way that strict differentiability and continuity are no longer required. This is precisely the motivation underlying the notion of a weak solution which has been developed by Lax [56, 86], Gel'fand [87], Oleinik [88], Rozhdestvensky [89], Kruzhkov [90], Kuznetsov [91], Douglis [92] and others. In the section that now follows an attempt will be made to give a brief introduction to the way in which such an extension may be achieved and the difficulties that ensue. For simplicity, the argument will be confined to a scalar equation, but the extension to a system follows without requiring any essentially new ideas.

For our starting point we take the equation

$$\frac{\partial u}{\partial t} + f(u)\frac{\partial u}{\partial x} = 0, \tag{4.35}$$

subject to the initial condition

$$u(x,0) = g(x), \tag{4.36}$$

and assume that f(u) is a continuous differentiable function of u. Then the first point to note is that (4.35) can be expressed in conservation form by defining

$$F(u) = \int f(u)du, \tag{4.37}$$

for then it becomes

$$\frac{\partial u}{\partial t} + \frac{\partial F}{\partial x} = 0. \tag{4.38}$$

Henceforth it will be with this equation, rather than with (4.35), that we shall work. In fact (4.35) was used on page 32 to illustrate the limitations of classical C^1 solutions and our conclusions in that context will also be of relevance here. Let us consider the half-plane $t > 0$ and recall that in general a unique solution to (4.35) and (4.36) will only exist for a finite time. Indeed, the envelope in Fig. 5 indicates the precise boundary of the region in which a unique C^1 solution exists for a particular choice of function f and initial condition g. As we have seen in Section 4.1 that conservation equations possess discontinuous solutions, or shocks, corresponding to a non-unique solution along an arc, we may conclude from this that the envelope in Fig. 5 represents the formation of a shock in the solution. Accordingly, and with reference now only to a general function f and initial condition g, let us consider some strip $0 < t < T$ in which the classical unique C^1 solution exists everywhere except on certain shock lines across which the solution is bounded. Adapting the notation of Section 4.1 we denote by u_- and u_+ the limiting values of u to the left and right of the shock under consideration, which from Theorem 4.2 are seen to vary continuously along the shock.

Then the bounded function u defined in the half plane $t > 0$ will be called a weak solution of (4.35) if in this half plane it satisfies the condition

$$\iint \left(\frac{\partial w}{\partial t} u + \frac{\partial w}{\partial x} F(u)\right) dxdt = 0, \tag{4.39}$$

for every twice continuously differentiable function $w(x,t)$ that vanishes outside some finite region in the half plane $t > 0$. Such functions w are called test functions [56, 93] and the closure of the region in which they are non-zero is then known as the support of the test functions. As a general classical C^1 solution to (4.35) subject to (4.36) has been found on page 32 we already know that if a weak solution satisfying (4.39) is also piecewise C^1, then it must be a classical solution wherever it is C^1. Thus a piecewise C^1 weak solution coincides with a piecewise C^1 classical solution, as would be expected of any reasonable extension of the concept of a solution.

Let us now show that there is a further common property shared between weak and piecewise C^1 classical solutions. This is that a piecewise C^1 weak solution satisfies the generalised Rankine-Hugoniot condition across a shock.

Consider the region R bounded by the closed arc ∂R and traversed by the line L across which a shock occurs. Denote the two sub-regions so defined by R_- and R_+ and their boundaries by ∂R_- and ∂R_+, and let the directed arcs along adjacent sides of L be ∂L_- and ∂L_+, as in Fig. 19.

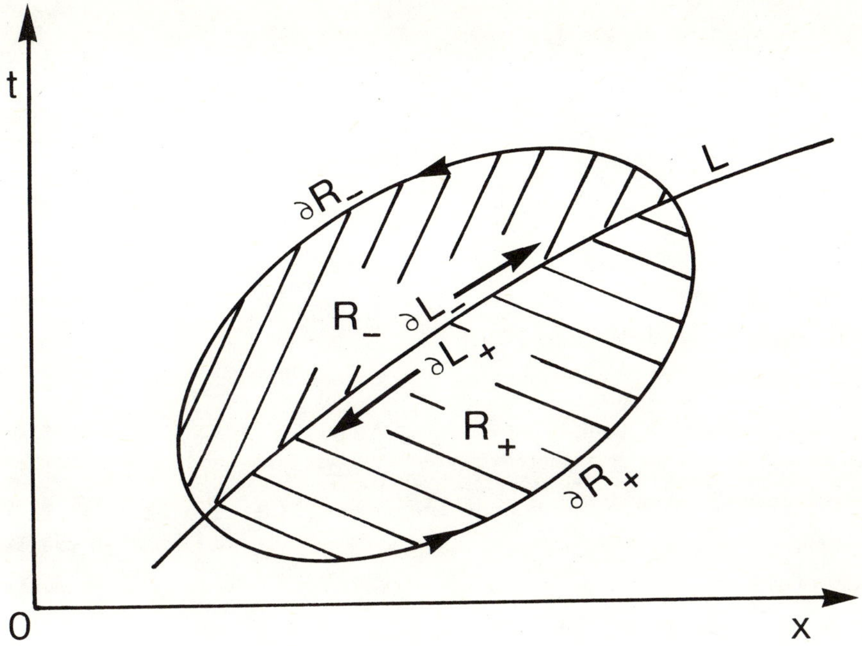

Fig. 19. Shock line L dividing R

Then $R = R_- \cup R_+$ and $\partial R = \partial R_- \cup \partial R_+$. The test functions w in (4.39) will be assumed to have their support in R so that the test functions w will vanish on ∂R. Thus (4.39) may be written

$$\iint_R \left(\frac{\partial w}{\partial t} u + \frac{\partial w}{\partial x} F(u)\right) dxdt = 0. \tag{4.40}$$

Now multiply (4.38) by w and integrate over R_- to obtain

$$\iint_{R_-} \left(w\frac{\partial u}{\partial t} + w\frac{\partial F}{\partial x}\right) dxdt = 0,$$

which may also be written in the form

$$\iint_{R_-} \left\{\frac{\partial(wu)}{\partial t} + \frac{\partial(wF)}{\partial x}\right\} dxdt - \iint_{R_-} \left\{\frac{\partial w}{\partial t} u + \frac{\partial w}{\partial x} F\right\} dxdt = 0. \tag{4.41}$$

Applying Green's theorem to the first terms in this result then transforms

(4.41) to

$$\oint_{\partial R_- \cup \partial L_-} -wFdt + wudx - \iint_{R_-} \left\{ \frac{\partial w}{\partial t} u + \frac{\partial w}{\partial x} F \right\} dxdt = 0. \qquad (4.42)$$

However as the support of the functions w lie in R, w will be zero on ∂R_- so that (4.42) reduces to

$$\oint_{\partial L_-} -wF(u_-)dt + w\, u_- dx - \iint_{R_-} \left\{ \frac{\partial w}{\partial t} u + \frac{\partial w}{\partial x} F \right\} dxdt = 0. \qquad (4.43)$$

A similar result applies with respect to R_+ where we find

$$\oint_{\partial L_+} -w\, F(U_+)dt + w\, u_+ dx - \iint_{R_+} \left\{ \frac{\partial w}{\partial t} u + \frac{\partial w}{\partial x} F \right\} dxdt = 0, \qquad (4.44)$$

the integration along ∂L_- and ∂L_+ being oppositely directed, as indicated in Fig. 19.

If (4.43) and (4.44) are now added, the sign of the line integral in (4.43) is reversed with a corresponding replacement of ∂L_- by ∂L_+ and result (4.40) is used we find

$$\oint_{\partial L_+} w\{(u_+ - u_-) \frac{dx}{dt} - (F(u_+) - F(u_-)) \}dt = 0, \qquad (4.45)$$

where as the point (x,t) is now constrained to lie on ∂L_+ the term (dx/dt) represents the speed of propagation $\tilde{\lambda}$ of the shock along L. As w is arbitrary, (4.45) can only be true if

$$\tilde{\lambda}(u_+ - u_-) = (F(u_+) - F(u_-)), \qquad (4.46)$$

which is the one dimensional form of the generalised Rankine-Hugoniot condition (4.14). This holds degenerately when u is continuous across L.

If, now, the support of w is allowed to be arbitrary, the same form of argument proves that piecewise C^1 solutions of (4.38) satisfying (4.46) across a shock will also be a weak solution of (4.38). We thus arrive at the following definition and theorem.

Definition 4.2 (Weak Solution)
The function u will be called a weak solution of

$$\frac{\partial u}{\partial t} + \frac{\partial F(u)}{\partial x} = 0$$

if for all twice continuously differentiable test functions w with support in $t > 0$ the function u is such that

$$\iint \left\{ \frac{\partial w}{\partial t} u + \frac{\partial w}{\partial x} F(u) \right\} dxdt = 0,$$

the integration being extended over the upper half plane $t > 0$. ■

Theorem 4.3 (Properties of Weak Solutions)
Let u be a weak solution of

$$\frac{\partial u}{\partial t} + \frac{\partial F(u)}{\partial x} = 0. \tag{4.47}$$

The following results are then true:

a) If u is piecewise C^1 in addition to being a weak solution it is also a piecewise C^1 classical solution of (4.47);

b) a piecewise C^1 weak solution of (4.47) satisfies the generalised Rankine-Hugoniot condition

$$\tilde{\lambda}(u_+ - u_-) = (F(u_+) - F(u_-)) \tag{4.48}$$

across a discontinuity moving with speed $\tilde{\lambda}$;

c) a necessary and sufficient condition for a piecewise C^1 classical solution of (4.47) to be a weak solution is that across a discontinuity moving with speed $\tilde{\lambda}$ it satisfies the generalised Rankine-Hugoniot condition (4.48). ■

The general objective when introducing a weak solution was to lift the requirements of strict continuity and differentiability that need to be imposed on classical solutions, since it is not usually known a priori how long they will remain C^1. In this respect the notion of a weak solution is successful and, furthermore, because of its method of definition the class of weak solutions is even wider than the class of piecewise C^1 functions so that considerable generality has been achieved. However, as we now show, this generality has been obtained at the cost of the uniqueness of a weak solution. More precisely, unlike a strict classical C^1 solution, a weak solution to (4.38) is not determined uniquely by the initial data (4.36). This is most easily demonstrated by means of a simple example.

Consider an equation of the form

$$\frac{\partial u}{\partial t} + \frac{\partial}{\partial x}(\tfrac{1}{3}u^3) = 0, \tag{4.48}$$

with

$$u(x,0) = \begin{cases} 0 & \text{for } x < 0 \\ 1 & \text{for } x > 0, \end{cases} \tag{4.49}$$

so that in (4.38) we have $F(u) = u^3/3$.

Then, as the equation is homogeneous, when it is differentiable a non-constant solution u will be a function of x/t, and it is easily verified that the function

$$u(x,t) = \begin{cases} 0 & \text{for } x/t < 0 \\ (x/t)^{\frac{1}{2}} & \text{for } 0 \leq x/t \leq 1 \\ 1 & \text{for } x/t > 1 \end{cases} \tag{4.50}$$

is a C^1 solution of (4.48) subject to the initial condition (4.49). This solution is continuous everywhere for $t > 0$, and it is differentiable everywhere except along each of the lines $x = 0$ and $x = t$ on which, due to the continuity of u, the generalised Rankine-Hugoniot condition holds in a degenerate form. It is a simple matter to verify directly that this piecewise C^1 classical solution is also a weak solution. The form of this solution is shown in Fig. 20.

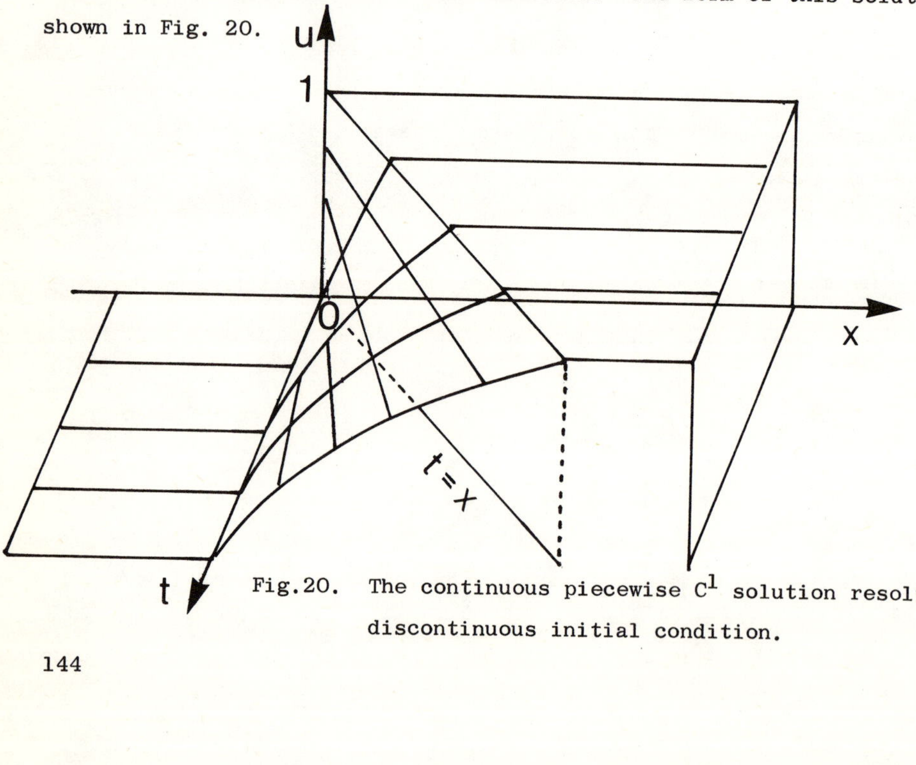

Fig.20. The continuous piecewise C^1 solution resolving a discontinuous initial condition.

Another weak solution follows by observing that a discontinuous function that is a C^1 solution away from the lines of discontinuity will be a weak solution provided the discontinuity satisfies the generalised Rankine-Hugoniot condition. Let us, then, seek an even simpler solution of the form

$$u(x,t) = \begin{cases} 0 \text{ for } x/t < k \\ 1 \text{ for } x/t > k, \end{cases}$$

by choosing k to satisfy the generalised Rankine-Hugoniot condition. Substitution into (4.46) coupled with the fact that the speed of shock propagation $\tilde{\lambda} = k$ then gives

$$k(1 - 0) = (\frac{1}{3} - 0),$$

or

$$k = 1/3.$$

The second weak solution satisfying (4.48) and (4.49) is thus

$$u(x,t) = \begin{cases} 0 \quad \text{for } x/t < 1/3 \\ 1 \quad \text{for } x/t > 1/3, \end{cases} \tag{4.51}$$

and this weak solution is piecewise constant, but is discontinuous across the line $3x = t$.

This justifies our earlier general assertion concerning the non-unique dependence of a weak solution on the initial data, and even when the initial data is smooth it is possible for weak solutions to be infinite in number. Such a property of weak solutions is not in accord with the situation in the physical world where a solution to a differential equation having physical significance is required to be unique. So, if this idea is to be of use with physical problems, some selection principle must be devised which will identify from amongst the class of weak solutions to a specific problem that one which has some additional property that makes it physically acceptable. Such a principle can be sought in various ways and it might, as in gas dynamics, be provided by the entropy condition mentioned on page 138 or, alternatively, it might be a mathematical selection process that yields a unique solution that can be given physical justification.

An early mathematical approach to this problem for the equation

$$\frac{\partial u}{\partial t} + u\frac{\partial u}{\partial x} = 0 \tag{4.52}$$

was proposed by Hopf [94] who began by considering the related equation

$$\frac{\partial u}{\partial t} + u\frac{\partial u}{\partial x} = \mu\frac{\partial^2 u}{\partial x^2} . \tag{4.53}$$

By means of a substitution he reduced (4.53) to the heat equation and was then able to deduce the asymptotic behaviour of the solution u_μ to (4.53) as $\mu \to 0$. This was of interest because, as $\mu \to 0$, so the parabolic equation (4.53) approaches the hyperbolic equation (4.52). He also established that the unique limiting solution of (4.53) was a weak solution of (4.52). The physical significance of this solution may be justified in the case of fluid dynamics by the fact that (4.53) was introduced by Burgers [95] as a model for the study of turbulence, in which the term μu_{xx} provides a measure of the dissipative viscous processes involved. The limit as $\mu \to 0$ may be considered to represent the approach of a viscous model to an inviscid model. A restricted case of this same problem was examined independently by Cole [96], who also reduced (4.53) to the heat equation in order to obtain a solution which he then used to study the competitive processes at work in the fluid. In another early paper Germain and Bader [97] succeeded in showing that, for given initial conditions, only one weak solution exists for (4.52) which also satisfies the generalised Rankine-Hugoniot condition, The essential details of this proof and a discussion of the fluid dynamic consequences are to be found in the review paper by Germain [85].

Hopf [98, 99] made further fundamental contributions to the development of selection principles which, by analogy with gas dynamics, are now known as "entropy conditions". In [99] he formulated a very general entropy condition which embraces both the one first developed by Oleinik [88] and the abstract one formulated by Kruzhkov [90]. The details of this work are beyond the scope of these notes but they may be found, together with further references,in the bibliographies of the works cited in this section. A very recent review of some of these ideas, and of shock waves in general, has been given by Lax [100]. This work also provides a useful introduction to the study of the decay of shock waves by Glimm and Lax [101]. The device of approximating (4.52) by (4.53) which contains a viscous-like term has also been utilised by other authors, both as an analytical tool and as a computational device to overcome the problem of handling a shock. See, for example, the papers by Oleinik [102] and by Conley and Smoller [103] for theoretical applications, and for numerical applications the papers by Lax [86] and Lax and Wendroff [104,105] and the books by Richtmyer and Morton [106] and by Ames [107].

Before closing this section we offer one more comment on properties of weak solutions, to the effect that they are not invariant under a change of dependent variable. To see this consider a C^1 solution to the conservation equation

$$\frac{\partial u}{\partial t} + \frac{\partial G}{\partial x} = 0$$

in some region of the (x,t)-plane. Then any C^1 function h(u) of u will satisfy the equation

$$\frac{\partial h}{\partial t} + G'(u) \frac{\partial h}{\partial x} = 0,$$

provided $h'(u) \neq 0$. This may in turn be expressed in the conservation form

$$\frac{\partial h}{\partial t} + \frac{\partial K}{\partial x} = 0$$

where $K'(u) = G'(u)\, h'(u)$ and a weak solution of the first conservation will not, in general, be a weak solution of the second. The shocks will also usually be different, for those of the first conservation equation will satisfy the generalised Rankine-Hugoniot condition

$$\tilde{\lambda}(u_+ - u_-) = (F(u_+) - F(u_-)),$$

while those for the second will obey the corresponding law

$$\tilde{\lambda}(h(u_+) - h(u_-)) = (K(u_+) - K(u_-)).$$

4.4 Conservation Equations with A Convex Extension

When the conservation system involved is symmetric hyperbolic, the ideas of Section 4.3 may be pursued in some detail without giving rise to undue difficulty. This we do now, basing our approach on the paper by Friedrichs and Lax [108] and on the contribution by Lax in the collection of papers edited by Zarantonello [109].

Consider a system of conservation equations

$$\frac{\partial U}{\partial t} + \frac{\partial G}{\partial x} = 0, \qquad (4.54)$$

with U and G = G(U) each n × 1 vectors and integrate it over an arbitrarily large interval $[-a,a]$ of the x-axis. Integrating the second term by parts then gives rise to the equation

$$\int_{-a}^{a} \frac{\partial U}{\partial t}\, dx + G\Big|_{a} - G\Big|_{-a} = 0.$$

Now for the class of solution vectors U that vanish sufficiently rapidly for large $|x|$, so that $G(\pm a, t) \to 0$ as $a \to \infty$, we see from the above result and the degenerate form of Theorem 4.1 that

$$\frac{d}{dt}\int_{-\infty}^{\infty} U dx = 0,$$

showing that the integral

$$\int_{-\infty}^{\infty} U dx$$

is a conserved quantity because it is independent of t.

The problem we now consider is, when is a new conservation system

$$\frac{\partial V}{\partial t} + \frac{\partial K}{\partial x} = 0, \tag{4.55}$$

with V, K functions of U, a direct consequence of the original law (4.54). To resolve this we need to make a direct comparison between (4.54) and (4.55) so that first we perform the indicated differentiations, when these equations become, respectively,

$$\frac{\partial U}{\partial t} + (\nabla_u G)\frac{\partial U}{\partial x} = 0 \tag{4.56}$$

and

$$(\nabla_u V)\frac{\partial U}{\partial t} + (\nabla_u K)\frac{\partial U}{\partial x} = 0. \tag{4.57}$$

Employing the summation convention, the j-th component of (4.56) may be written

$$\frac{\partial u_j}{\partial t} + \frac{\partial g_j}{\partial u_\ell}\frac{\partial u_\ell}{\partial x} = 0, \tag{4.58}$$

while equation (4.57) itself becomes

$$\frac{\partial V}{\partial u_j}\frac{\partial u_j}{\partial t} + \frac{\partial K}{\partial u_\ell}\frac{\partial u_\ell}{\partial x} = 0. \tag{4.59}$$

Consequently, comparing (4.58) and (4.59), we conclude that (4.57) will be a consequence of (4.56) only if

$$\frac{\partial V}{\partial u_j}\frac{\partial g_j}{\partial u_\ell} = \frac{\partial K}{\partial u_\ell} . \tag{4.60}$$

Let us now assume that this condition is true, and differentiate it with respect to u_h, when we find

$$\frac{\partial g_j}{\partial u_\ell}\left[\frac{\partial}{\partial u_h}\left\{\frac{\partial V}{\partial u_j}\right\}\right] + \frac{\partial V}{\partial u_j}\left\{\frac{\partial^2 g_j}{\partial u_h \partial u_\ell}\right\} = \frac{\partial^2 K}{\partial u_\ell \partial u_h} .$$

The second term on the left hand side and the right hand side are both symmetric in ℓ and h, so that the first term must also be symmetric. We have thus shown that if (4.60) is true, then

$$\frac{\partial g_j}{\partial u_\ell}\left[\frac{\partial}{\partial u_h}\left\{\frac{\partial V}{\partial u_j}\right\}\right] = \frac{\partial g_j}{\partial u_h}\left[\frac{\partial}{\partial u_\ell}\left\{\frac{\partial V}{\partial u_j}\right\}\right] . \tag{4.61}$$

If, now, we multiply (4.58) by $\partial^2 V/\partial u_j \partial u_h$ and sum with respect to j we find

$$\frac{\partial^2 V}{\partial u_j \partial u_h}\frac{\partial u_j}{\partial t} + \frac{\partial^2 V}{\partial u_j \partial u_h}\frac{\partial g_j}{\partial u_\ell}\frac{\partial u_\ell}{\partial x} = 0. \tag{4.62}$$

This will be equivalent to (4.54) if the matrix $\{\partial^2 V/\partial u_j \partial u_h\}$ is non-singular, and we here take note of the fact that system (4.62) is symmetric. Hence, whenever (4.54) is hyperbolic, and (4.60) is true, the equivalent system (4.62) will be symmetric hyperbolic. The results quoted on page 49 are thus applicable and show that a unique solution will exist in some neighbourhood of the initial data.

As the hyperbolicity of (4.62) implies that the matrix $\{\partial^2 V/\partial u_j \partial u_h\}$ is positive definite we may assert that V is a convex function of the elements u_h [17] and so arrive at the following conclusion.

Theorem 4.4 (Uniqueness Theorem)

If the system of conservation equations (4.54) is such that it implies a new conservation equation (4.55) with the property that the new conserved quantity V is a convex function of the original elements $u_1, u_2, \ldots, u_n$ of U, then the initial value problem for (4.54) has a unique solution in the neighbourhood of the initial time. ∎

For the derivation of an entropy principle from these arguments and its relation to the one by Kruzhkov [90] we refer to the source references [108, 109].

4.5 Evolutionary Condition for Shocks in Hyperbolic Systems of Conservation Type.

We now return to the idea of a weak solution in connection with the use of piecewise C^1 solutions to represent one dimensional shocks. As already remarked, weak solutions are not unique, and if they are to be used to describe shocks in relation to a physical problem, then some way must be found to eliminate the non-physical shock solutions from consideration. In this problem which is more restricted than the one that was considered in Section 4.3 we shall appeal to what has become called the evolutionary condition [8]. This is the requirement that a physically realisable shock must be stable to small disturbances which, it will be recalled, was one of Hadamard's requirements for a well-posed problem.

Let us now outline the nature of this problem by considering in one-dimension the interaction of a plane shock wave and a small plane disturbance wave. For simplicity we assume that this shock wave separates two slightly disturbed adjacent regions in each of which the solution depends only on x and t. The shock will then occur across a plane normal to the x-axis, as will the small disturbance waves. Since it is postulated that only a small disturbance is involved, and the undisturbed solution will be assumed to be known, we may consider the linearised equations governing the propagation of the disturbances adjacent to the shock, while on the shock itself the disturbances must satisfy the linearised version of the generalised Rankine-Hugoniot conditions. The linearised systems of equations on either side of the shock will differ according to the given initial solutions U_o and U_1 on opposite sides of the shock with respect to which linearisation has taken place.

Because of linearity, it is possible to resolve any small plane disturbance wave into a sum of plane waves, each associated with one of the characteristics of the system and having a specific amplitude, care being taken to distinguish between waves that approach the shock and waves that leave it. The amplitude of the incident waves will be determined by the initial conditions, and, on meeting the shock, waves will be both reflected and transmitted. The amplitudes of these induced waves will be determined by the conditions at the shock and by the incident waves. Whenever, with a given incident wave, the shock is such that the reflected and transmitted waves are uniquely determined, then the shock will be said to be evolutionary in nature. In the event that no solution exists, or that it is not unique, the shock will be said to be non-evolutionary. Indeed, a generalisation of this important problem will be one that will be examined in detail in Chapter 5 from the point of view of the transmitted and reflected waves and their propagation characteristics.

Having now outlined what is to follow let us consider the details, of finding a condition for evolutionarity, and in doing so follow in the main the method devised by Jeffrey and Taniuti [8], while incorporating a modification of the argument due to Collins [14] that has been attributed by him to D.S. Butler.

Consider a system of conservation laws

$$\frac{\partial F}{\partial t} + \frac{\partial G}{\partial x} = 0 \tag{4.63}$$

where $F = F(U)$, $G = G(U)$ are $n \times 1$ vectors of the $n \times 1$ vector U. The performing the indicated differentiations gives

$$(\nabla_u F)\frac{\partial U}{\partial t} + (\nabla_u G)\frac{\partial U}{\partial x} = 0,$$

so that writing $A = C^{-1}B$, with $B = (\nabla_u G)$ and $C = (\nabla_u F)$, we arrive at the equivalent quasilinear system of n equations

$$\frac{\partial U}{\partial t} + A\frac{\partial U}{\partial x} = 0 \tag{4.64}$$

and, because we assume (4.63) to be hyperbolic, the matrix A will have real eigenvalues and a full set of eigenvectors which span the space E_n. For this exposition we shall assume (4.63) to be totally hyperbolic and hence the n eigenvalues of A will be distinct.

When weak solutions of (4.63) occur which are discontinuous they will satisfy the generalised Rankine-Hugoniot condition

$$\tilde{\lambda}(F(U_+) - F(U_-)) = (G(U_+) - G(U_-)) . \tag{4.65}$$

Since we shall only consider the case in which the solutions $U_{\pm}$ adjacent to the discontinuity are constant vectors it will be convenient to adopt a coordinate system that moves with the shock. Accordingly, we set

$$x' = x - \tilde{\lambda}t \text{ and } t' = t \tag{4.66}$$

when (4.63) becomes

$$\frac{\partial F}{\partial t'} + \frac{\partial}{\partial x'}(G - \tilde{\lambda}F) = 0$$

and (4.64) itself becomes

$$\frac{\partial U}{\partial t'} + (A - \tilde{\lambda}I)\frac{\partial U}{\partial x'} = 0.$$

Henceforth we shall omit the prime on the variable t' and suppose that the discontinuity occurs across the origin $x' = 0$. In keeping with the problem outlined at the outset we now set the solutions to the left and right of $x' = 0$ equal to $U^{(0)}$ and $U^{(1)}$, and define a discontinuous function $\bar{U}$ by writing

$$\bar{U} = \begin{cases} U^{(0)} & \text{for } x' < 0 \\ U^{(1)} & \text{for } x' > 0. \end{cases}$$

Setting $\bar{F} = F(\bar{U})$ and $\bar{G} = G(\bar{U})$ and employing the notation $[\cdot]$ used to denote a jump quantity in (4.13) we know that across the shock

$$\tilde{\lambda}[\bar{F}] = [\bar{G}].$$

The function $\bar{U}$ is a weak solution and we now subject it to a small perturbation δU, so that the disturbed solution is

$$U = \bar{U} + \delta U,$$

and the perturbed shock speed becomes $\tilde{\lambda} + \delta\tilde{\lambda}$.

Performing the linearisation mentioned at the outset then yields the following condition for δU

$$\frac{\partial(\delta U)}{\partial t} + (\bar{A} - \tilde{\lambda}I)\frac{\partial(\delta U)}{\partial x'} = 0, \tag{4.67}$$

where $\bar{A} = A(\bar{U})$.

Correspondingly, the condition at the shock becomes

$$\delta\tilde{\lambda}[F] = [G - \tilde{\lambda}F],$$

where F, G are here determined appropriate to the argument $\bar{U} + \delta U$. The linearised version of this result at x' = 0 is easily seen to be

$$[(\bar{B} - \tilde{\lambda}\bar{c})\delta U] = \delta\tilde{\lambda}[\bar{F}]. \tag{4.68}$$

Employing the linear superposition principle that is now applicable, and the fact that a disturbance associated with the eigenvalue $\lambda^{(i)}$ of A must be proportional to the corresponding i-th right eigenvector r_i of A, permits us to express $\delta U^{(0)}$ and $\delta U^{(1)}$ as follows:

<u>disturbance to the left</u>

$$\delta U^{(0)} = \sum_{k=1}^{n} r_k^{(0)} f_k^{(0)}\left(t - \frac{x'}{\lambda_k^{(0)}}\right), \tag{4.69a}$$

<u>disturbance to the right</u>

$$\delta U^{(1)} = \sum_{k=1}^{n} r_k^{(1)} f_k^{(1)}\left(t - \frac{x'}{\lambda_k^{(1)}}\right), \tag{4.69b}$$

where the superscript (0) refers to the solution vector $U^{(0)}$ and the superscript (1) to the solution vector $U^{(1)}$. The functions $f_k^{(0)}$ and $f_k^{(1)}$ are scalar functions of their arguments and characterise the wave profile of the linear disturbance waves associated with the k-th characteristic corresponding to the eigenvalue λ_k on either side of the shock. If $\tau_k^{(j)}$ are the eigenvalues of $A^{(j)}$, for j = 0,1, and $\lambda_k^{(j)}$ are the eigenvalues of $\bar{A}$, both arranged in increasing order of magnitude, then $\tau_k^{(j)} = \tilde{\lambda} + \lambda_k^{(j)}$ for k = 1,2,...,n.

If the shock is to be evolutionary, we require that when at t = 0 the incident disturbance is specified, the resulting disturbances must remain small and be uniquely determined for all subsequent times $t > 0$. The initial data then corresponds to the specification of all the functions $f_k^{(j)}(-x'/\lambda_k^{(j)})$. When $\lambda_k^{(0)} > 0$, then since the corresponding function $f_k^{(0)}(-x'/\lambda_k^{(0)})$ is defined for $x' < 0$, the function $f_k^{(0)}(t - x'/\lambda_k^{(0)})$ is known for $t > 0$ and $x' < 0$ from the given initial values. Correspondingly, if $\lambda_k^{(1)} < 0$, the function $f_k^{(1)}(t - x'/\lambda_k^{(1)})$ is known for $t > 0$ and $x' > 0$ from the given initial values. These functions describe the incident waves. However, if $\lambda_k^{(0)} < 0$ and $\lambda_k^{(1)} > 0$, the corresponding functions $f_k^{(0)}(t - x'/\lambda_k^{(0)})$ and $f_k^{(1)}(t - x'/\lambda_k^{(1)})$ are only known for negative values of their arguments, and these then represent outgoing waves. In the event that an eigenvalue $\lambda_k^{(j)}$ is zero, the corresponding term in (4.69a,b) will be replaced by $r_k^{(j)}f_k^{(j)}(x')$, which then represents a stationary wave. As $f_k^{(j)}(x')$ is known for $t = 0$, it is thus known for all $t > 0$.

Let us now suppose that $\lambda_k^{(0)} < 0$ for $k = 1,2,\ldots,p$ and that $\lambda_k^{(1)} > 0$ for $k = m+1, m+2,\ldots,n$. Then the p functions $f_k^{(0)}(t - x'/\lambda_k^{(0)})$ and the n-m functions $f_k^{(1)}(t - x'/\lambda_k^{(1)})$ are unknown for positive values of their arguments The remaining functions of each set of n functions will, however, be known for the entire range of their arguments.

The boundary condition (4.68) at $x' = 0$ may be written

$$(B^{(1)} - \tilde{\lambda}C^{(1)})\delta U^{(1)} - (B^{(0)} - \tilde{\lambda}C^{(0)})\delta U^{(0)} = \delta\tilde{\lambda}(F^{(1)} - F^{(0)}),$$

so that on substituting for $\delta U^{(j)}$ from (4.69a,b) we find

$$\sum_{k=1}^{n}\lambda_k^{(1)}C^{(1)}r_k^{(1)}f_k^{(1)}(t) - \sum_{k=1}^{n}\lambda_k^{(0)}C^{(0)}r_k^{(0)}f^{(0)}(t)$$

$$= \delta\tilde{\lambda}(t)(F^{(1)} - F^{(0)}) \text{ for } t \geq 0. \qquad (4.70)$$

However, as the functions $f_k^{(0)}(t)$ for $k = p+1, p+2,\ldots,n$ and $f_k^{(1)}(t)$ for $k = 1,2,\ldots,m$ are known, these equations comprise a set of n equations for the p unknown functions $f_k^{(0)}(t)$ for $k = 1,2,\ldots,p$ and the n-m functions $f_k^{(1)}(t)$ for $k = m+1,m+2,\ldots,n$ and the shock speed perturbation $\delta\tilde{\lambda}(t)$. These equations will have a unique and non-trivial solution if $m = p+1$ and the vectors $C^{(1)}r_k^{(1)}$ corresponding to positive $\lambda_k^{(1)}$ and the vector $F^{(1)} - F^{(0)}$ are linearly independent. This follows for there will then be n linearly independent equations from which to determine n-1 unknown functions $f_k^{(j)}$ and the shock speed perturbation $\delta\tilde{\lambda}$. Under these circumstances we may determine the unknown functions $f_k^{(j)}(t)$ and hence determine the disturbance wave profiles $f_k^{(j)}(t - x'/\lambda_k^{(j)})$ corresponding to outgoing waves for positive values of their arguments. These functions are then determined for all $t > 0$ and for $x' < 0$ when $j = 0$ and for $x' > 0$ when $j = 1$.

We thus arrive at the conclusion that the problem for (4.67) is well-posed if there are only n-1 functions $f_k^{(j)}(t - x'/\lambda_k^{(j)})$ corresponding to outgoing waves. Expressed differently, (4.67) is well-posed if the total number of negative eigenvalues $\lambda_k^{(0)}$ and positive eigenvalues $\lambda_k^{(1)}$ is n-1, and if the vectors $C^{(j)}r_k^{(j)}$ corresponding to these outgoing waves and the vector $[\bar{F}]$ are linearly independent. Finally, we require the functions $f_k^{(j)}$ to be small so that the linearisation remains valid.

The requirement that there be only n-1 outgoing waves enables us to deduce the condition found by Lax [56] that a shock should be evolutionary. Whenever a shock is evolutionary there will be n-1 non-zero eigenvalues corresponding to outgoing waves, where k-1 will be associated with negative eigenvalues $\lambda_1^{(0)}, \lambda_2^{(0)}, \ldots, \lambda_{k-1}^{(0)}$, and n-k with positive eigenvalues $\lambda_{k+1}^{(1)}, \lambda_{k+2}^{(1)}, \ldots, \lambda_n^{(1)}$, for some index k, which may take any of the values k = 1,2, ...,n. Hence we arrive at the following definition and theorem.

<u>Definition 4.3</u> (Evolutionary Shock)

A shock will be said to be evolutionary if the resolution of the interaction of an incident small disturbance wave and the shock into outgoing disturbance waves that remain small and a disturbed shock motion comprise a well-posed problem. ▮

<u>Theorem 4.5</u> (Evolutionary Condition)

i) A shock is evolutionary if the inequalities

$$\lambda_{k-1}^{(0)} < 0 \leq \lambda_k^{(0)} \text{ and } \lambda_k^{(1)} \leq 0 < \lambda_{k+1}^{(1)}$$

are satisfied for some integer k.

ii) The undisturbed shock speed $\tilde{\lambda}$ is related to the eigenvalue $\lambda_p^{(j)}$ of $\bar{A}$ and to the eigenvalue $\tau_p^{(j)}$ of $A^{(j)}$ by the equation

$$\lambda_p^{(j)} = \tau_p^{(j)} - \tilde{\lambda} .$$

iii) For some integer k the shock speed $\tilde{\lambda}$ must satisfy the inequalities

$$\tau_{k-1}^{(0)} < \tilde{\lambda} \leq \tau_k^{(0)} \text{ and } \tau_k^{(1)} \leq \tilde{\lambda} < \tau_{k+1}^{(1)} .$$

The integer k is called the index of the shock. ▮

Our conclusions now enable us to re-phrase Definition 4.3 and also to give other definitions of importance which are in agreement with the corresponding concepts in fluid dynamics.

<u>Definition 4.4</u> (Shock Classification)

(i) A shock with speed $\tilde{\lambda}$ which is different from any of the eigenvalues $\lambda_k^{(j)}$ on either side of the shock and which satisfies inequalities (iii) of Theorem 4.5 will be called a genuine or evolutionary shock.

(ii) A degenerate shock with speed $\tilde{\lambda}$ that coincides with an eigenvalue

$\lambda_k^{(j)}$ on either side of the shock will be called a contact discontinuity.

(iii) A shock with speed $\tilde{\lambda}$ that coincides with an eigenvalue $\lambda_k^{(j)}$ on only one side of the shock will be called an intermediate discontinuity. ∎

The ideas of this section are easily illustrated by means of a simple application to one-dimensional gas dynamics in which there are three different types of possible discontinuity. These are a shock moving to the left, a contact discontinuity and a shock moving to the right. Using the information from page 100 together with the notation of this section we know that, when arranged in order of magnitude, the appropriate eigenvalues are

$$\lambda_1^{(j)} = u^{(j)} - a^{(j)}, \quad \lambda_2^{(j)} = u^{(j)}, \quad \lambda_3^{(j)} = u^{(j)} + a^{(j)},$$

for $j = 0,1$. As there are three characteristics involved, for a shock to be genuine there must only be two outgoing waves.

Let us now compute the number of outgoing waves under different circumstances so that we may determine when a shock will be genuine, or evolutionary, to give it the alternative name. To do this we consider, relative to the shock between regions (0) to the left and (1) to the right, when the waves are outgoing. For region (0) this corresponds to negative eigenvalues or speeds while for region (1) this corresponds to positive ones. As we shall need to consider the speeds in regions (0) and (1) it will be helpful to use the sound speeds $a^{(0)}$ and $a^{(1)}$ to sub-divide the speed ranges.

Region (0) with $0 < u^{(0)} < a^{(0)}$

one outgoing wave to the left with speed $\lambda_1^{(0)} = u^{(0)} - a^{(0)} < 0$.

Region (0) with $u^{(0)} > a^{(0)}$

no outgoing waves to the left

Region (1) with $0 < u^{(1)} < a^{(1)}$

two outgoing waves to the right with speeds $\lambda_2^{(1)} = u^{(1)} > 0$ and $\lambda_3^{(1)} = u^{(1)} + a^{(1)} > 0$.

Region (1) with $u^{(1)} > a^{(1)}$

three outgoing waves to the right with speeds $\lambda_1^{(1)} = u^{(1)} - a^{(1)} > 0$, $\lambda_2^{(1)} = u^{(1)} > 0$ and $\lambda_3^{(1)} = u^{(1)} + a^{(1)} > 0$.

A diagramatic plot of this situation is shown in Fig. 21 in which the number in a region indicates the number of outgoing waves in that region. Only in the shaded region is the number of outgoing waves equal to two. The shaded region is thus the only one for which genuine shocks are possible.

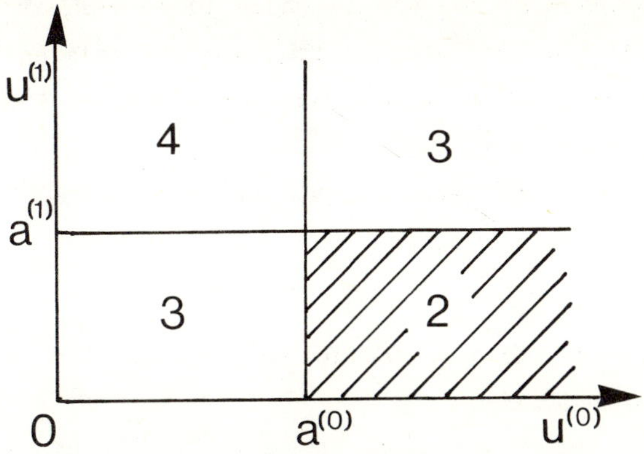

Fig. 21. The integer indicates the number of outgoing waves in the region.

Interpreted in terms of a shock moving to the right we have

$$u^{(0)} < \tilde{\lambda} \leq u^{(0)} + a^{(0)} \text{ and } u^{(1)} + a^{(1)} \leq \tilde{\lambda}.$$

These inequalities show that the shock speed $\tilde{\lambda}$ exceeds the gas speed on either side of the shock. Also, as particles cross the shock from right to left, they show that the shock is supersonic with respect to the gas in region (1) and subsonic with respect to the gas in region (0). This is in agreement with the results obtained when the entropy condition from thermodynamics is used to select the physically realisable gas shock [9]. Thus, when applied to a gas shock, the evolutionary condition selects the same solution as the thermodynamic entropy condition mentioned on page 138. In this sense the evolutionary condition contains the thermodynamic entropy condition. The restriction made here to constant solutions $U^{(0)}$ and $U^{(1)}$ on adjacent sides of the shock may be lifted and the results of our theorems shown still to apply in the general case.

4.6 Connection of Solutions by k-Shocks

In the previous section a precise condition was given in Theorem 4.5 for an evolutionary, or genuine, shock in terms of the eigenvalues. These eigenvalues were arranged in order of magnitude and indexed so that the evolutionary shock condition was associated with a particular index k, where $k = 1,2,\ldots,n$. There are, thus, n different shocks and to distinguish between them we shall refer to a k-shock with the understanding that k is the index of the condition in Theorem 4.5 which is satisfied by the shock under consideration. Let us now examine the way in which these k-shocks may be used to join a constant solution $U^{(0)}$ to all the possible adjacent solutions $U^{(1)}$, and in doing this we follow the fundamental paper by Lax [56] and the account given by Jeffrey and Taniuti [8].

The conservation system with which we shall work is

$$\frac{\partial U}{\partial t} + \frac{\partial G}{\partial x} = 0, \tag{4.71}$$

where U and $G = G(U)$ are $n \times 1$ vectors, and for this system the generalised Rankine-Hugoniot condition becomes

$$\tilde{\lambda}[U] = [G]. \tag{4.72}$$

So, when $\tilde{\lambda}$ is eliminated, this represents n-1 equations for the jump quantities and so, as required by the evolutionary condition, there must be n-1 characteristics radiating out from the shock. The precise problem we now investigate is the nature of the set of all solutions $U^{(1)}$ to which $U^{(0)}$ may be connected by means of a k-shock on the right.

First we observe that the solutions $U^{(1)}$ are not arbitrary since they must conform to both the generalised Rankine-Hugoniot condition (4.72) and the condition of Theorem 4.5. So, if $U^{(0)}$ is kept fixed, then since (4.72) with $\tilde{\lambda}$ eliminated comprise n-1 conditions connecting $U^{(0)}$ and $U^{(1)}$ we see that $U^{(1)}$ must form a one-parameter family of the form

$$U^{(1)} = U(\epsilon) \text{ with } U(0) = U^{(0)}.$$

The shock speed $\tilde{\lambda}$ will also be a function of ϵ so that

$$\tilde{\lambda} = \tilde{\lambda}(\epsilon),$$

where ϵ may, perhaps, be taken as the jump $[u_m]$ of the m-th element of U. Our object will now be to determine the behaviour of ϵ across a k-shock so that we may infer from it the behaviour of the one parameter family $U^{(1)}$. Towards this end we now determine $\lambda(0)$, $\tilde{\lambda}(0)$, $\dot{\lambda}(0)$ and $\dot{\tilde{\lambda}}(0)$, where a dot signifies differentiation with respect to ϵ.

Differentiating (4.72) with respect to ϵ and setting $\epsilon = 0$ gives

$$\tilde{\lambda}(0)\dot{U} = A\dot{U} \quad \text{for} \quad \epsilon = 0,$$

where $A = (\nabla_u G)$.

This may be re-written

$$(A - \tilde{\lambda}(0)I)\dot{U} = 0 \quad \text{for} \quad \epsilon = 0$$

showing that

$$\tilde{\lambda}(0) = \lambda_k(U^{(0)}) \qquad (4.73)$$

and

$$\dot{U}(0) = \alpha r^{(k)}(U^{(0)}), \qquad (4.74)$$

where α is a scalar multiplier and $r^{(k)}$ is the k-th right eigenvector of A corresponding to λ_k. Hereafter the subscript k, the index of the shock, will be omitted and a suitable choice of ϵ will enable us to set $\alpha \equiv 1$ so that we have

$$\dot{U}(0) = r(U^{(0)}), \qquad (4.75)$$

where the normalisation of r will be determined later.

Differentiating (4.72) twice with respect to ϵ and using equations (4.73) to (4.75) at $\epsilon = 0$ then brings us to the result

$$\lambda\ddot{U} + 2\dot{\tilde{\lambda}}r = A\ddot{U} + \dot{A}r. \qquad (4.76)$$

If, now, we differentiate the defining relation $Ar = \lambda r$ for the right eigenvector r once with respect to ϵ we find

$$A\dot{r} + \dot{A}r = \lambda\dot{r} + \dot{\lambda}r, \qquad (4.77)$$

so that multiplying (4.76) and (4.77) on the left by the left eigenvector ℓ corresponding to λ and subtracting gives for $\epsilon = 0$,

$$2\dot{\tilde{\lambda}}(0) = \dot{\lambda}(0). \qquad (4.78)$$

Now subtracting (4.77) from (4.76) gives

$$\lambda(\ddot{U} - \dot{r}) = A(\ddot{U} - \dot{r})$$

showing that the vector $\ddot{U} - \dot{r}$ must be proportional to the right eigenvector r.

Thus we may write

$$\ddot{U}(0) = \dot{r} + \beta r$$

with β a scalar constant. The constant β may be made zero by a suitable choice of parameterisation which we now assume to be made. We thus have the result

$$\ddot{U}(0) = \dot{r} = (\nabla_u r)\dot{U} = (\nabla_u r)r. \tag{4.79}$$

At this point we need to assume that the λ_k characteristic field is genuinely nonlinear so that

$$(\nabla_u \lambda_k) r^{(k)} \neq 0,$$

or, since the index k is omitted elsewhere, that

$$(\nabla_u \lambda) r \neq 0.$$

This being so we may normalise r so that

$$(\nabla_u \lambda) r = 1. \tag{4.80}$$

Equation (4.75) then shows that

$$1 = (\nabla_u \lambda(0)) \dot{U}(0) = \dot{\lambda}(0)$$

So that with the aid of (4.78) we conclude

$$\dot{\lambda}(0) = 1 \text{ and } \dot{\tilde{\lambda}}(0) = \tfrac{1}{2}. \tag{4.81}$$

From Theorem 4.5 we now deduce that ϵ must be negative in order that the discontinuity involved is a k-shock. The normalisation of the parameterisation is provided by equations (4.75), (4.79), and (4.80).

Equation (4.78) may be interpreted as implying that the shock speed, up to terms of order ϵ^2, is the arithmetic mean of the characteristic speeds to the right and left of the shock.

There is a connection with the problem that gave rise to Fig. 20 on page 144 that must be explored here. Such a solution, which involves the resolution of a discontinuous piecewise constant initial condition, was first studied by Riemann and for obvious reasons the C^1 non-constant part of the solution is called a simple wave centred on the origin.

Returning for a moment to the work of Section 3.4, we now consider a simple wave centred on the origin which depends only on the ratio x/t. Let us assume λ_k is constant and that $(\nabla_u \lambda_k) r^{(k)}$ and $r^{(k)}$ may be normalised by requiring

$$(\nabla_u \lambda_k) r^{(k)} = 1. \tag{4.82}$$

Then as the solution U in the non-constant differentiable region is taken to be centred we have

$$U(x,t) = V(x/t),$$

where V must be determined by the equations

$$J_s^{(k)}(V) = \text{const. for } s = 1,2,\ldots,n-1 \tag{4.83}$$

for the (n-1) generalised λ_k-Riemann invariants associated with $r^{(k)}$. It then follows from our knowledge of generalised simple waves that the lines x/t = ξ = const are characteristics of the system. Consequently V satisfies the equation

$$\lambda_k(V(\xi)) = \xi. \tag{4.84}$$

The two equations (4.83) and (4.84) now comprise a system of n equations for the elements $v_1, v_2, \ldots, v_n$ of V. These will have a unique solution because of the normalisation used for $r^{(k)}$ in the condition (4.82).

If the region of the centred simple wave is given by $a < x/t < b$ and it is connected with the constant solutions $U^{(1)}$ and $U^{(0)}$ for $x/t \geq b$ and $x/t \leq a$, respectively, then the two constant states must have the restriction that they have the same k-th generalised Riemann invariants with

$$\lambda_k(U^{(0)}) < \lambda_k(U^{(1)}).$$

Thus the two constant solutions $U^{(0)}$ and $U^{(1)}$ are connected via a k-th centred simple wave and across this wave the jumps of the (n-1)-th generalised Riemann invariants J are zero, so that

$$[J] = 0, \tag{4.85}$$

provided

$$\lambda(U^{(0)}) < \lambda(U^{(1)}). \tag{4.86}$$

The two states may be given by $U^{(1)} = U(\epsilon)$ and $U^{(0)} = U(0)$ where now ϵ is a parameter representing the strength of the simple wave. From (4.85) and the differentiation of $J(U(\epsilon))$ with respect to ϵ at $\epsilon = 0$ we arrive at the result

$$(\nabla_u J)\dot{U} = 0.$$

So, from the definition of a generalised Riemann invariant (3.49) we conclude

$$\dot{U}(0) = r(U^{(0)}), \tag{4.87}$$

with the parameterisation now fixed so that there is a unit constant of proportionality. The normalisation condition (4.80) then shows that

$$\dot{\lambda} = (\nabla_u \lambda)\dot{U} = (\nabla_u \lambda)\ r = 1, \tag{4.88}$$

which shows that λ increases with ϵ.

So, from (4.86), we see that $\epsilon > 0$. If, now, we differentiate (4.85) twice with respect to ϵ at $\epsilon = 0$ and then use (4.87) we find

$$(\nabla_u J)\ddot{U} + \frac{d}{d\epsilon}(\nabla_u J)r = 0. \tag{4.89}$$

However, differentiating equation (3.49) with respect to ϵ and setting $\epsilon = 0$ gives

$$(\nabla_u J)\dot{r} + \frac{d}{d\epsilon}(\nabla_u J)r = 0 \tag{4.90}$$

from which, by subtraction, we find

$$(\nabla_u J)(\ddot{U} - \dot{r}) = 0$$

or, the equivalent result,

$$\ddot{U} = \dot{r} = (\nabla_u r)\dot{U} = (\nabla_u r)r. \tag{4.91}$$

Equations (4.75), (4.87), (4.79) and (4.91) show that $\dot{U}(0)$ and $\ddot{U}(0)$ are the same for the k-th centred simple wave and the k-shock. As the k-th generalised Riemann invariants do not change across a k-th centred simple wave we conclude that the change in a k-th generalised Riemann invariant across a k-shock is of third order in ϵ. The conclusions of this section may now be expressed as a theorem.

Theorem 4.6 (Properties of k-Shocks)

i) The shock speed, up to terms of order ϵ^2, is the arithmetic mean of the characteristic speeds to the left and right of the shock.

ii) The change in a k-th generalised Riemann invariant across a k-shock is of third order in ϵ. $\blacksquare$

The problem on page 144 for a scalar equation that gave rise to Fig. 20 was an example of the so-called Riemann problem which is an initial value problem for (4.71) of the form

$$U(x,0) = \begin{cases} U^{(0)} & \text{for } x < 0 \\ U^{(n)} & \text{for } x > 0. \end{cases} \tag{4.92}$$

This involves the resolution of an initial discontinuity in the solution vector, and from what has been developed so far we conclude that when a continuous and differentiable non-constant solution is involved then the evolutionary condition requires it to be a centred simple wave. In general, the solution to system (4.71) subject to (4.72) will be a combination of genuine k-shocks, centred simple waves, intermediate and exceptional discontinuities, all centred on the origin.

These waves divide the space into n+1 constant states $U^{(0)}, U^{(1)}, \ldots, U^{(n)}$, when $U^{(k)}$ and $U^{(k+1)}$ are separated by a genuine k-shock or centred simple wave of the k-th type or by an exceptional discontinuity if the k-th characteristic field is exceptional.

As $U^{(k+1)}$ can be expressed in terms of a parameter ϵ_{k+1} and the vector $U^{(k)}$, and this form of relationship exists between any two adjacent solutions, we arrive at the results

$$U^{(n)} = U^{(n)}(U^{(0)};\ \epsilon_1, \epsilon_2, \ldots, \epsilon_n)$$

$$U^{(0)} = U^{(0)}(U^{(0)};\ 0,\ 0, \ldots, 0),$$

where ϵ_{i+1} is the parameter relating $U^{(i+1)}$ and $U^{(i)}$. These form n inhomogeneous equations for the n parameters $\epsilon_1, \epsilon_2, \ldots, \epsilon_n$ once $U^{(0)}$ and $U^{(n)}$ have been specified. However, we have

$$\dot{U}(0) = r(U^{(0)}) \text{ and } \frac{\partial U^{(n)}}{\partial \epsilon_k} \propto r^{(k)}$$

for $\epsilon_1 = \epsilon_2 = \ldots = \epsilon_n = 0$, so that $\partial U^{(n)}/\partial\epsilon_1$, $\partial U^{(n)}/\partial\epsilon_2, \ldots, \partial U^{(n)}/\partial\epsilon_n$ are linearly independent at the origin in the ϵ-space. The implicit function theorem then assures us of the existence of a unique solution in the neighbourhood of the origin. We thus have our last theorem.

<u>Theorem 4.7</u> (Existence Theorem for Riemann Problem)
There exists a neighbourhood of $U^{(0)}$ such that if $U^{(n)}$ belongs to this neighbourhood, the Riemann problem (4.71) subject to (4.92) has a solution. The solution comprises n+1 constant solutions connected by centred simple waves, and the solution is unique provided the intermediate states lie in a suitable neighbourhood of $U^{(0)}$. ▌

A considerable amount of literature exists concerning the decay of shocks and, from the point of view of continuum mechanics, much of this has been reviewed by Sedov [110]. A detailed study has been made of various mathematical aspects of the problem by various authors including Glimm [111], Conway and Smoller [112], Glimm and Lax [101] and Lax [100]. Many other applications have been made and from these we mention only the application to gas dynamics by Burnside and Mackie [113], to magnetohydrodynamics by Gunderson [114], to solid mechanics by Inan [115] and to the decay of weak electromagnetic shocks by Jeffrey and Korobeinikov [142].

5 Development of shocks from Lipschitz continuous data

5.1 C^n Discontinuities and Wavefront Propagation in One Space Dimension and Time

It has been demonstrated in Chapters 1 and 2 that in one space dimension and time a characteristic curve may be defined as any curve $\mathscr{C}$ in the (x,t)-plane across which the solution vector U is continuous but its derivative normal to $\mathscr{C}$ is indeterminate. This was, indeed, the property that enabled us in Fig. 1. to identify a wavefront associated with a hyperbolic system as a line on a continuous solution surface across which there is a bounded discontinuous change in the first derivative of U normal to $\mathscr{C}$. Such a wavefront manifests itself as a "crease" in the solution surface, and because it arises from a Lipschitz discontinuity in the first derivative of U normal to $\mathscr{C}$ it will be called a C^1 discontinuity.

An obvious generalisation is to describe a continuous vector function U with a bounded discontinuity in the n-th derivative of U normal to $\mathscr{C}$ as a C^n discontinuity. It follows from Section 2.4 that C^n discontinuities also propagate along characteristics and so can be identified with a line on the solution surface whose projection onto the (x,t)-plane is the characteristic, or wavefront trace, $\mathscr{C}$. If, in addition to a C^n discontinuity, there is also a C^1 discontinuity propagating along $\mathscr{C}$, then there will be a visible wavefront on the solution surface in the form of a "crease" as in Fig.1. If, however, no C^1 discontinuity exists, then although a wavefront still exists it will be associated with a smooth solution surface; that is, with a solution surface for which at least the first order partial derivatives are continuous. It is natural to refer to such a wave as a smooth fronted wave, and later we provide an elementary example of a smooth fronted C^2 wave in connection with the shallow water wave approximation.

We will show, as would be expected, that the propagation characteristics of a C^n discontinuity are determined in terms of the propagation characteristics of all the associated C^r discontinuities, for $r = 1,2,\ldots,n-1$. When only the behaviour on the wavefront is required, as is often the case with physical problems, attention may be confined to the wavefront trace $\mathscr{C}$.

The behaviour of the propagating C^n discontinuity along $\mathscr{C}$ may then be determined in terms of a system ordinary differential equations with the elapsed time as parameter along $\mathscr{C}$. Providing the details of this analysis will be one of the main tasks in this present chapter, and arising from it will be the associated question of determining when and where a wavefront steepens to the point at which a shock forms on it. That is, when and where on $\mathscr{C}$ a discontinuity in the solution vector U first forms. Thus, unlike the analysis of Section 3.5 which examined the possibility of shock formation on a global basis for the solution to a reducible system, the analysis of this chapter is local to the wavefront trace. It is, however, still very general since it embraces inhomogeneous hyperbolic systems with n dependent variables and, when they are of conservation form, it allows for systems with discontinuous coefficient matrices.

The study of hyperbolic systems with discontinuous coefficient matrices is important since they occur frequently in physical problems, usually being associated with a change of material properties or a change of geometrical configuration.

When such systems occur the coefficient matrices will be discontinuous across some line $\mathscr{D}$ in the (x,t)-plane, and we shall then assume that the systems are of conservation form. This will enable us to confine the notion of a weak solution to a piecewise smooth solution with a jump, or shock, across $\mathscr{D}$ that is completely characterised by a generalised Rankine-Hugoniot type algebraic jump condition. For the solution to be physically meaningful, in the sense that it is unique and depends continuously on its initial data, it is also necessary to require that it satisfies the evolutionary condition of Chapter 4. If the coefficient matrices are not discontinuous there will be no need to restrict consideration to a conservation system.

Also of concern to us will be the manner of interaction between a propagating C^1 discontinuity and an established shock across the discontinuity line $\mathscr{D}$. This will be seen to give rise to reflected and transmitted waves, the precise analysis of this being given in the general case in Section 5.5 and for some important special cases in Section 5.7. The time and place of shock formation on a wavefront due to the steepening of a C^1 discontinuity will be discussed in Section 5.6, and this may either take place before $\mathscr{D}$ has been reached or after it has been passed.

The method of analysis used in this chapter finds its origin in references [8,116], and it has been refined and extended subsequently in a number of other papers, to some of which reference will be made later. From amongst this more recent work we mention the two papers by Jeffrey [117,118], since it is upon these that the present account is largely based. In that work, because only C^1 discontinuities were involved, the term weak discontinuity was employed in place of C^1 discontinuity, while the term strong discontinuity was used for a shock.

Because of the simplicity of structure of the system involved, the detailed application of the work of this chapter has been confined to the shallow water wave model. This same model has been used both with C^1 discontinuities in Section 5.9, and in the extension to C^2 discontinuities in Section 5.10. Other applications of a more complex nature exist in the literature of which we mention only the applications to continuum mechanics by Jeffrey and Teymur [62], Jeffrey and Inan [119] and Suhubi and Jeffrey [120].

Closely related to the C^1 discontinuity propagation problem that arises out of this analysis when the coefficient matrices are continuous in the paper by John [66] to which reference has already been made. Apart from contributing to the general understanding of processes that lead to the development of singularities, this paper also examines plane wave propagation in elasticity and shows that, generally, a small finite disturbance of compact support when superimposed on a constant state of strain will result in a motion that becomes singular in a finite elapsed time. Another contribution to this general subject, in the context of gas dynamics, was made in the early paper by Nitsche [68] which was first mentioned in Section 3.5. He looked at both the general problem of shock formation and at the time of existence of a solution to a hyperbolic system with given initial data from which certain types of wavefront analysis arise as special cases. The paper by Meyer [121] also contributes to a gas dynamic problem involving shock formation in flow through a duct, though with the exception of the work by John the methods used by these authors are not closely related to those employed here.

5.2 Conservation System With Discontinuous Coefficients

As the situation across all discontinuity lines is the same, it will suffice to examine the consequences when a single discontinuity line $\mathscr{D}$ occurs in the solution, and to this effect we consider two first order quasilinear systems, each of n equations, of the form

$$U_t + AU_x + B = 0 \tag{5.1}$$

and

$$U^*_t + A^* U^*_x + B^* = 0, \tag{5.2}$$

with the arbitrary differentiable initial conditions $U(x,0) = \Phi(x)$, $U^*(x,0) = \Phi^*(x)$ defined, respectively, on adjacent intervals $\mathscr{I}_u$, $\mathscr{I}_{u*}$ of the initial line

$$\mathscr{I}_u = \{x | a \leqq x < x_d;\ t = 0\} \text{ and } \mathscr{I}_{u*} = \{x | x_d < x \leqq b;\ t = 0\}. \tag{5.3}$$

Later, in connection with equation (5.4), we shall use the fact that for systems (5.1) and (5.2) of conservation type the jump discontinuity in Φ and Φ^* across the point $(x_d, 0)$ is not arbitrary.

In all discussions that follow it will be assumed that the systems (5.1) and (5.2) are hyperbolic, and that the numbers a and b have been chosen such that the analysis of the propagation of a weak discontinuity starting at $(x_o, 0)$ in the domain of dependence $\mathscr{I} = \mathscr{I}_u \cup \mathscr{I}_{u*}$, is confined to the domain of determinacy $\mathscr{R}$ associated with $\mathscr{I}$ and illustrated in Fig. 22.

The symbols U, U^*, B and B^* denote column vectors each with n components u_i, u^*_i, b_i and b^*_i respectively, whilst A and A^* denote square matrices of order n with A a function of U, x and t and A^* a function of U^*, x and t. To allow maximum generality we also assume that $B = B(U,x,t)$ and $B^* = B^*(U^*,x,t)$. All vectors and matrices will be assumed to be real and continuously differentiable functions of U, U^*, x and t in their respective domains of determinacy within $\mathscr{R}$, but to be discontinuous across some curve $\mathscr{D}$ that divides $\mathscr{R}$. Specifically, it will be assumed that in addition to the propagating C^1 or weak discontinuity which starts at a point $(x_o, 0) \in \mathscr{I}$, and forms the wavefront under discussion, the solution also has an initial strong discontinuity to the right of $(x_o, 0)$ at the point $(x_d, 0) \in \mathscr{I}$ through which the curve $\mathscr{D}$ followed by the strong discontinuity enters the upper half plane. To simplify later arguments it will be convenient to denote the determinacy domains to the left and right of $\mathscr{D}$, corresponding to $\mathscr{I}_u$ and $\mathscr{I}_{u*}$, by $\mathscr{R}_u$ and $\mathscr{R}_{u*}$, respectively, so that $\mathscr{R} = \mathscr{R}_u \cup \mathscr{R}_{u*}$, as shown in Fig. 22.

We now assume that systems (5.1) and (5.2) are of conservation type, so that they may be written in the form first shown in equation (1.11). Then,

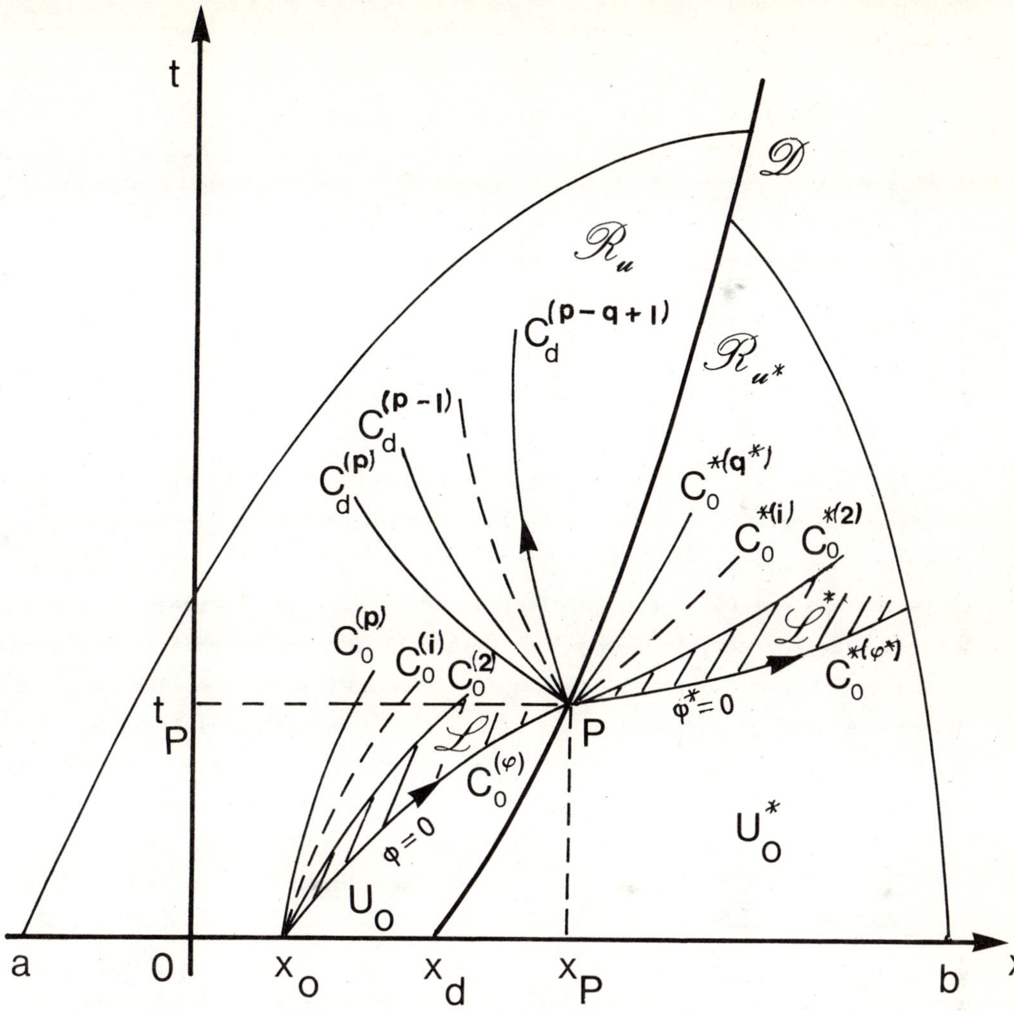

Fig. 22. Wavefront traces, discontinuity line $\mathscr{D}$ and domains of dependence and determinacy.

in accordance with Chapter 4, it is appropriate to refer to the strong discontinuity across $\mathscr{D}$ as a shock and the line itself as the shock line. The piecewise smooth weak solutions that comprise such shocks are completely characterised by requiring that across $\mathscr{D}$ a Rankine-Hugoniot type condition of the form of (4.12) is satisfied.

That is, we use an algebraic jump condition to relate solutions U and U^* on the adjacent sides $\mathscr{D}_-$ and $\mathscr{D}_+$ of $\mathscr{D}$ which lie, respectively, to the left and right of the shock line. When the one-dimensional systems of equations (5.1) and (5.2) are of conservation type they may, respectively, be written in the forms

$$F_t + \text{div } G = H \quad \text{and } F^*_t + \text{div } G^* = H^*,$$

where the known vectors F, G and H depend on U and the known vectors F^*, G^* and H^* depend on U^*. The jump conditions then take the form of the generalized Rankine-Hugoniot relations of fluid mechanics

$$\{G\}_{\mathcal{d}^-} - \{G^*\}_{\mathcal{d}^+} = \tilde{\lambda}\{F\}_{\mathcal{d}^-} - \tilde{\lambda}\{F^*\}_{\mathcal{d}^+}. \tag{5.4}$$

Here $\{\cdot\}_{\mathcal{d}^\pm}$ signifies that the associated quantity $\{\cdot\}$ is to be evaluated on the sides $\mathcal{D}_\pm$ of $\mathcal{D}$, whilst $\tilde{\lambda}$ denotes the speed of propagation of the shock that exists across $\mathcal{D}$. To select the physically relevant piecewise smooth solution across $\mathcal{D}$ we also assume that the evolutionary condition has been employed in conjunction with (5.4).

As the systems (5.1) and (5.2) are hyperbolic, the matrices A, A^* will each possess n real eigenvalues together with full sets of n linearly independent left eigenvectors. Suppose that there are p distinct eigenvalues $\lambda^{(i)}$ of A and p* distinct eigenvalues $\lambda^{*(i)}$ of A^*, and that the multiplicities of $\lambda^{(i)}$ and $\lambda^{*(i)}$ are constant, and equal to m_i and m^*_i, respectively. Then it must follow that

$$\sum_{i=1}^{p} m_i = \sum_{i=1}^{p^*} m^*_i = n, \tag{5.5}$$

and if $\ell^{(i,k)}$, $\ell^{*(i,k)}$ are the left eigenvectors of A and A^* corresponding, respectively, to the eigenvalues $\lambda^{(i)}$ and $\lambda^{*(i)}$, we have

$$\ell^{(i,k)} A = \lambda^{(i)}\ell^{(i,k)}, \quad k = 1,2,\ldots,m_i \tag{5.6}$$

and

$$\ell^{*(i,k)} A^* = \lambda^{*(i)}\ell^{*(i,k)}, \quad k = 1,2,\ldots,m^*_i. \tag{5.7}$$

The characteristic forms of equations (5.1) and (5.2) which will be required subsequently are obtained by pre-multiplication of those equations by $\ell^{(i,k)}, \ell^{*(i,k)}$ to obtain, respectively,

$$\ell^{(i,k)}U_t + \lambda^{(i)}\ell^{(i,k)}U_x + b^{(i,k)} = 0 \tag{5.8}$$

for $i = 1,2,\ldots, p$ and $k = 1,2,\ldots,m_i$ with $b^{(i,k)} = \ell^{(i,k)} B$, and

$$\ell^{*(i,k)}U_t^* + \lambda^{*(i)}\ell^{*(i,k)}U_x^* + b^{*(i,k)} = 0 \tag{5.9}$$

for $i = 1,2,\ldots,p^*$ and $k = 1,2,\ldots,m_i^*$ with $b^{*(i,k)} = \ell^{*(i,k)}B^*$. There are p distinct families of characteristic curves $C^{(i)}$ in the determinacy domain $\mathscr{R}_u$ defined by the equations

$$C^{(i)} : \frac{dx}{dt} = \lambda^{(i)}, \quad i = 1,2,\ldots,p, \tag{5.10}$$

and p^* distinct families of characteristic curves $C^{*(i)}$ in the determinacy domain $\mathscr{R}_{u*}$ defined by the equations

$$C^{*(i)} : \frac{dx}{dt} = \lambda^{*(i)}, \; i = 1,2,\ldots,p^*. \tag{5.11}$$

For any given value of i, the m_i equations (5.8) are valid along characteristic curves $C^{(i)}$, and the m_i^* equations (5.9) are valid along characteristic curves $C^{*(i)}$.

As the projection onto the (x,t)-plane of the wavefront defined by a propagating C^1 discontinuity in U or U^* will lie along a characteristic curve which we call the wavefront trace it will be useful to introduce equations for these particular characteristics in $\mathscr{R}_u$ and $\mathscr{R}_{u*}$. In particular, we denote the equation of the characteristic in $\mathscr{R}_u$ through $(x_o,0)$ forming the initial wavefront trace by

$$\varphi(x,t) = 0, \tag{5.12}$$

and the equation of the new wavefront trace through P, the point where $\varphi(x,t) = 0$ meets $\mathscr{D}$ (Fig. 22) by

$$\varphi^*(x,t) = 0. \tag{5.13}$$

For convenience, the initial value problem that will concern us will be stated in terms of the fastest C^1 discontinuity in U that propagates initially to the right in the determinacy domain $\mathscr{R}_u$, and moves along the characteristic $C_o^{(1)}$ with equation $\varphi(x,t) = 0$ which passes through the point $(x_o,0)$. For this purpose we shall suppose the $\lambda^{(i)}$ to be ordered in magnitude with $\lambda^{(i)}$ corresponding to the wavefront trace and $\lambda^{(p)}$ to the slowest moving wave in $\mathscr{R}_u$. To emphasize that $\lambda^{(1)}$ describes the wavefront trace $\varphi = 0$ we set

$\lambda^{(1)} \equiv \lambda^{(\varphi)}$ and write $C_o^{(1)} \equiv C_o^{(\varphi)}$, so that the characteristic $C_o^{(\varphi)}$ is determined as the solution of

$$C^{(\varphi)} : \frac{dx}{dt} = \lambda^{(\varphi)}, \tag{5.14}$$

which passes through the point $(x_o, 0)$.

Similarly, if $\varphi^*(x,t) = 0$ is the equation of the fastest C^1 discontinuity in U^* moving to the right in $\mathscr{R}_{u*}$ along the characteristic $C_o^{*(\varphi*)}$, then using the same system of ordering the $\lambda^{*(i)}$ as was used with the $\lambda^{(i)}$, and setting $C_o^{*(\varphi*)} \equiv C_o^{*(1)}$, we see that $C_o^{*(\varphi*)}$ is determined as the solution of

$$C^{*(\varphi*)} : \frac{dx}{dt} = \lambda^{*(\varphi*)}, \tag{5.15}$$

which passes through the point (x_p, t_p). Here the point (x_p, t_p) is the point P at which $C_o^{(\varphi)}$ intersects $\mathscr{D}$. The problem now is to determine the propagation law for the C^1 discontinuity in U that exists across $C_o^{(\varphi)}$, the effect the shock at P on $\mathscr{D}$ has on the transmitted C^1 discontinuity in U^* along $C_o^{*(\varphi*)}$ and along the other characteristics $C_o^{*(i)}$ in $\mathscr{R}_{u*}$, the nature of the reflected C^1 discontinuities in $\mathscr{R}_u$ and, when it exists, the point (x_c, t_c) on $C_o^{*(\varphi*)}$ at which a shock first forms in the solution U^* within $\mathscr{R}_{u*}$.

5.3 Change of Coordinates and Jump Conditions

To be consistent, the solution vector $U(x,t) = U_o(x,t)$ to equation (5.1) in $\mathscr{R}_u$, ahead of the wavefront trace $C_o^{(\varphi)}$ and behind $\mathscr{D}$, must be such that

$$U_{ot} + A_o U_{ox} + B_o = 0, \tag{5.16}$$

where in (5.16) the suffix zero is used to signify that the associated quantity refers to the solution in $\mathscr{R}_u$ ahead of $C_o^{(\varphi)}$. Here, on account of the fact that the C^1 discontinuity is propagating to the right, the phrase "ahead of the wavefront trace" implies to the right of it and, conversely, "behind the wavefront trace" implies to its left. As in Chapter 2 and in previous work [8,116,117], in region $\mathscr{R}_u$ we introduce new semi-curvilinear coordinates φ, t' in place of the independent variables x, t in such a manner that the wavefront trace $C_o^{(\varphi)}$ is embedded as one of the family of coordinate lines. The construction of such a semi-characteristic coordinate system may be achieved by requiring φ, t' to satisfy the equations

$$\varphi_t + \lambda^{(\varphi)} \varphi_x = 0 \text{ and } t' = t. \tag{5.17}$$

That the new coordinates have the desired property follows, as in Chapters 2 and 3, from the fact that along φ = constant

$$\varphi_t + \frac{dx}{dt}\varphi_x = 0, \tag{5.18}$$

for then comparison with equation (5.17) shows that

$$\frac{dx}{dt} = \lambda^{(\varphi)} \quad \text{along } \varphi = \text{constant in } \mathscr{R}_u. \tag{5.19}$$

As the coordinate variable φ has only been introduced as the solution to the differential equation displayed in (5.17), it is necessary to supplement this equation by specifying an initial condition for φ along $t = 0$. This may be arbitrary, provided only that φ has the monotone property of a coordinate variable, and that $\varphi = 0$ corresponds to the characteristic $C_o^{(\varphi)}$ through the point $(x_o,0)$.

We may achieve this most simply by setting

$$\varphi(x,0) = x - x_o, \tag{5.20}$$

for then φ has the required coordinate property and $\varphi > 0$ ahead of the wavefront trace $C_o^{(\varphi)}$ on which $\varphi = 0$, and $\varphi < 0$ behind $C_o^{(\varphi)}$.

The Jacobian of transformation (5.17) is

$$x_\varphi = 1/\varphi_x, \tag{5.21}$$

and as condition (5.20) for φ implies that $x_\varphi = 1$ along the initial line, it follows from continuity arguments that the Jacobian will be non-vanishing for at least a finite elapsed time after $t = 0$. Thus the mapping in $\mathscr{R}_u$ from x,t to φ,t' will be one-one, and coordinate lines of the same family will not intersect, thereby bringing about non-uniqueness in the transformation, provided $x_\varphi \neq 0$.

Similarly, in region $\mathscr{R}_{u*}$, we introduce semi-characteristic coordinates by requiring that we change from the independent variables x,t to φ^*, t^* through the equations

$$\varphi^*_t + \lambda^{*(\varphi*)}\varphi^*_x = 0 \quad \text{and } t^* = t. \tag{5.22}$$

Here, in place of condition (5.20) along the initial line $t = 0$, we now impose a condition along the line $t = t_p$ of the form

$$\varphi^*(x,t_p) = x - x_p. \tag{5.23}$$

This ensures that, as required, $\varphi^* = 0$ along $C_o^{*(\varphi*)}$, with $\varphi^* > 0$ ahead of $C_o^{*(\varphi*)}$ and $\varphi^* < 0$ behind $C_o^{*(\varphi*)}$ in $\mathscr{R}_{u*}$. The Jacobian in $\mathscr{R}_{u*}$ is

$$x_{\varphi*} = 1/\varphi^*_x, \tag{5.24}$$

and the remarks concerning the non-vanishing of x_φ in $\mathscr{R}_u$ apply equally to the non-vanishing of $x_{\varphi*}$ in $\mathscr{R}_{u*}$.

In terms of these new variables we have:

$$\underline{\text{in } \mathscr{R}_u} \quad U_t = \varphi_t U_\varphi + U_{t'}, \quad U_x = \varphi_x U_\varphi, \text{ and} \tag{5.25}$$

$$\underline{\text{in } \mathscr{R}_{u*}} \quad U^*_t = \varphi^*_t U_{\varphi*} + U^*_{t*}, \quad U^*_x = \varphi^*_x U^*_{\varphi*}. \tag{5.26}$$

Employing these results in equations (5.8) and (5.9) together with (5.14) (5.15), (5.21) and (5.24) we eventually arrive at the following results:

$\underline{\text{in } \mathscr{R}_u}$ $\underline{\text{behind } C_o^{(\varphi)}}$

$$\ell^{(i,k)}\left\{x_\varphi \frac{\partial}{\partial t'} + (\lambda^{(i)} - \lambda^{(\varphi)})\frac{\partial}{\partial \varphi}\right\}U + x_\varphi b^{(i,k)} = 0, \tag{5.27}$$

for $i = 1,2,\ldots,p$ and $k = 1,2,\ldots,m_i$, and

$\underline{\text{in } \mathscr{R}_{u*}}$ $\underline{\text{behind } C_o^{*(\varphi*)}}$

$$\ell^{*(i,k)}\left\{x_{\varphi*} \frac{\partial}{\partial t*} + (\lambda^{*(i)} - \lambda^{*(\varphi*)})\frac{\partial}{\partial \varphi*}\right\}U^* + x_{\varphi*} b^{*(i,k)} = 0, \tag{5.28}$$

for $i = 1,2,\ldots,p^*$ and $k = 1,2,\ldots,m^*_i$.

To proceed further we now need to recognize that in $\mathscr{R}_u$ the operation $\partial/\partial\varphi$ represents differentiation normal to $C_o^{(\varphi)}$, whilst $\partial/\partial t'$ represents differentiation parallel to $C_o^{(\varphi)}$. As the wavefront trace $C_o^{(\varphi)}$ is the projection onto the (x,t)-plane of the C^1 discontinuity in U, it follows at once that U_φ is discontinuous across $C_o^{(\varphi)}$, whilst U and $U_{t'}$ are continuous across it. Similarly, $U^*_{\varphi*}$ is discontinuous across $C_o^{*(\varphi*)}$, whilst U^* and U^*_{t*} are continuous across it. Thus we are led to define the following C^1 discontinuity jump quantities:

$\underline{\text{in } \mathscr{R}_u}$ $\underline{\text{across } C_o^{(\varphi)}}$

U is continuous: $\quad [U]_{\varphi=0_+}^{\varphi=0_-} = 0$

$U_{t'}$ is continuous: $\qquad [U_{t'}]^{\varphi=0_-}_{\varphi=0_+} = 0$

U_φ is discontinuous: $\qquad [U_\varphi]^{\varphi=0_-}_{\varphi=0_+} = \Pi(t') \neq 0$

x_φ is discontinuous: $\qquad [x_\varphi]^{\varphi=0_-}_{\varphi=0_+} = X(t') \neq 0,$

in $\mathscr{R}_{u*}$ across $C_o^{*(\varphi*)}$

U^* is continuous: $\qquad [U^*]^{\varphi^*=0_-}_{\varphi^*=0_+} = 0$

U^*_{t*} is continuous: $\qquad [U^*_{t*}]^{\varphi^*=0_-}_{\varphi^*=0_+} = 0$

$U^*_{\varphi*}$ is discontinuous: $\qquad [U^*_{\varphi*}]^{\varphi^*=0_-}_{\varphi^*=0_+} = \Pi^*(t^*) \neq 0$

$x_{\varphi*}$ is discontinuous: $\qquad [x_{\varphi*}]^{\varphi^*=0_-}_{\varphi^*=0_+} = X^*(t^*) \neq 0,$

where the jump in a scalar or vector quantity α across $C^{(\varphi)}$ is denoted by $[\alpha]^{\varphi=0_-}_{\varphi=0_+} \equiv \alpha\big|_{\varphi=0_-} - \alpha\big|_{\varphi=0_+}$, and the jump quantity $[\alpha^*]^{\varphi^*=0_-}_{\varphi^*=0_+}$ across $C_o^{*(\varphi*)}$ is similarly defined.

It follows from the coordinate transformations used that $t_\varphi = t_{\varphi*} \equiv 0$, so that no jump quantities need be defined with respect to either of these quantities.

Analogous to equation (5.16), the solution vector $U^*(x,t) = U^*_o(x,t)$ to equation (5.2) in $\mathscr{R}_{u*}$, ahead of $\mathscr{D}$ and ahead of the wavefront trace $C^{*(\varphi*)}$ must be such that

$$U^*_{ot} + A^*_o U^*_{ox} + B^*_o = 0, \tag{5.29}$$

where again the suffix zero is used to signify that the associated quantity is to be evaluated in this region.

We observe here that jumps in the vectors U_φ and $U^*_{\varphi*}$ are possible across each characteristic in $\mathscr{R}_u$ and $\mathscr{R}_{u*}$, respectively. Hence, when differencing an equation or expression across a wave front trace, as will be done in the next section, it is important to consider only the result obtained by differencing across the one characteristic curve represented by the wavefront trace itself, and not across any additional characteristics entering $\mathscr{R}_u$

from$(x_o,0)$. In the case of $\mathcal{R}_u$ this is equivalent to interpreting the region behind the wavefront trace $C_o^{(\varphi)}$ as the region $\mathcal{L}$ shown in Fig.22, lying behind $C_o^{(\varphi)}$ but ahead of $C_o^{(2)}$, the next characteristic issuing out into $\mathcal{R}_u$ from the point $(x_o,0)$. A similar argument applies to the jump in $U^*_{\varphi*}$, when a corresponding region $\mathcal{L}^*$ may be defined relative to $C_o^{*(\varphi*)}$ and $C_o^{*(2)}$ at the point P in $\mathcal{R}_{u*}$.

The effect of differencing across several characteristic curves issuing from a point will be considered later in Section 5.4, in connection with the resolution of the initial C^1 discontinuity present across $\varphi^*=0$ in $\mathcal{R}_{u*}$ after the C^1 discontinuity vector propagating along $\varphi=0$ has passed $\mathcal{D}$ at P.

5.4 Transport Equations for C^1 Discontinuities

We now derive the transport equations for the C^1 discontinuities in U and U^* as they move, respectively, along $C_o^{(\varphi)}$ and $C_o^{*(\varphi*)}$. Equation (5.27) is valid behind $C_o^{(\varphi)}$ in $\mathcal{R}_u$, and it follows from (5.16) and the arguments that led to (5.27) that immediately ahead of $C_o^{(\varphi)}$ we also have the equation

$$\ell_o^{(i,k)}\left\{ x_{\varphi o}\frac{\partial}{\partial t'} + (\lambda_o^{(i)} - \lambda_o^{(\varphi)})\frac{\partial}{\partial\varphi}\right\}U_o + x_{\varphi o}b_o^{(i,k)} = 0, \tag{5.30}$$

for $i = 1,2,\ldots,p$ and $k = 1,2,\ldots,m_i$, where $x_{\varphi o}$ signifies the quantity x_φ immediately ahead of $C_o^{(\varphi)}$. Now, since A, B are assumed to be continuous functions of their arguments and U is continuous across $C_o^{(\varphi)}$, it follows at once that $\ell^{(i,k)}$ will also be continuous across $C_o^{(\varphi)}$. Hence, differencing equations (5.27) and (5.30) across the single characteristic $C_o^{(\varphi)}$, assuming $\lambda^{(i)} \neq \lambda^{(\varphi)}$ and using the jump relations just defined we arrive at the $n-m_1$ equations

$$\ell_o^{(i,k)}U_{ot'}X + (\lambda_o^{(i)} - \lambda_o^{(\varphi)})\,\ell_o^{(i,k)}\Pi + b_o^{(i,k)}X = 0, \tag{5.31}$$

for $i = 2,3,\ldots,p$ and $k = 1,2,\ldots,m_i$.

Eliminating $U_{ot'}$ between equations (5.30) and (5.31) then gives the result

$$-\ell_o^{(i,k)}U_{\varphi o}X + \ell_o^{(i,k)}\Pi x_{\varphi o} = 0$$

or, finally,

$$-\ell_o^{(i,k)}U_{ox}X + \ell_o^{(i,k)}\Pi = 0, \tag{5.32}$$

for $i = 2,3,\ldots,p$ and $k = 1,2,\ldots,m_i$.

To obtain a further $m_1 + 1$ equations connecting X and Π we now set $\lambda^{(i)} = \lambda^{(\varphi)}$ in equations (5.27) and (5.30), differentiate the resulting equations with respect to φ, and then difference them in turn across the single characteristic $C_o^{(\varphi)}$ making use of the jump relations. In the case of (5.27), setting $\lambda^{(i)} = \lambda^{(\varphi)}$ (i.e., i = 1) gives

$$\ell^{(\varphi,k)} U_{t'} + b^{(\varphi,k)} = 0$$

for $k = 1,2,\ldots,m_1$, when differentiation with respect to φ allowing for the fact that $\ell^{(\varphi,k)}$, $b^{(\varphi,k)}$ depend explicitly on U, x and t then yields

$$[(\nabla_u \ell^{(\varphi,k)}) U_\varphi]' \, U_{t'} + \ell_x^{(\varphi,k)} x_\varphi U_{t'} + \ell_t^{(\varphi,k)} t_\varphi U_{t'} + \ell^{(\varphi,k)} U_{\varphi t'}$$
$$+ (\nabla_u b^{(\varphi,k)}) U_\varphi + b_x^{(\varphi,k)} x_\varphi + b_t^{(\varphi,k)} t_\varphi = 0 \qquad (5.33)$$

for $k = 1,2,\ldots,m_1$, where, as usual, ∇_u denotes the gradient operator with respect to the dependent variables $u_1,u_2,\ldots,u_n$ and the prime associated with a matrix is used to denote the matrix transpose operation. In the coordinate transformation used $t_\varphi \equiv 0$, so that subtracting (5.33) from the equivalent equation derived from (5.30), which holds ahead of $C_o^{(\varphi)}$, and using the jump relations gives the required m_1 equations

$$[(\nabla_u \ell^{(\varphi,k)})_o \Pi]' \, U_{ot'} + \ell_{ox}^{(\varphi,k)} X U_{ot'} + \ell_o^{(\varphi,k)} \Pi_{t'}$$
$$+ (\nabla_u \ell^{(\varphi,k)})_o \Pi + b_{ox}^{(\varphi,k)} X = 0, \qquad (5.34)$$

for $k = 1,2,\ldots,m_1$.

Equations (5.32) and (5.34) together provide n equations for the n components $\pi_1,\pi_2,\ldots,\pi_n$ of the column vector Π, so that one further equation is still required connecting the jump discontinuity vector Π and the scalar discontinuity quantity X. This is obtained by observing that behind $C_o^{(\varphi)}$ in $\mathscr{L}$, and ahead of $C_o^{(2)}$, it follows from (5.10) with i = 1 that

$$C^{(\varphi)} : \frac{dx}{dt} = \lambda^{(\varphi)}$$

or, equivalently,

$$\frac{\partial x}{\partial t'} = \lambda^{(\varphi)}.$$

Hence differentiation with respect to φ, allowing for explicit dependence on x and t, yields

$$\frac{\partial}{\partial t'}(x_\varphi) = (\nabla_u \lambda^{(\varphi)})U_\varphi + \lambda_x^{(\varphi)} x_\varphi + \lambda_t^{(\varphi)} t_\varphi \tag{5.35}$$

immediately behind $C_o^{(\varphi)}$, and an equivalent expression exists ahead of $C_o^{(\varphi)}$. Differencing these equations across the single characteristic $C_o^{(\varphi)}$, and again using the jump conditions together with the result $t_\varphi \equiv 0$, now gives

$$\frac{dX}{dt'} = (\nabla_u \lambda^{(\varphi)})_o \Pi + \lambda_{ox}^{(\varphi)} X. \tag{5.36}$$

Between equations (5.32), (5.34) and (5.36) we have n + 1 equations from which the n elements $\pi_1, \pi_2, \ldots, \pi_n$ of the vector Π and the scalar quantity X may be determined. These equations are called the transport equations for the C^1 discontinuity in $\mathscr{R}_u$.

Exactly analogous arguments give for the corresponding transport equations in $\mathscr{R}_{u*}$:

$$-\ell_o^{*(i,k)} U_{ox}^* X^* + \ell_o^{*(i,k)} \Pi^* = 0 \tag{5.37}$$

for $i = 2,3,\ldots,p*$ and $k = 1,2,\ldots,m_i^*$,

$$[(\nabla_{u*}\ell^{*(\varphi*,k)})_o \Pi^*]' U_{ot*}^* + \ell_{ox}^{*(\varphi*,k)} X^* U_{ot*}^* + \ell^{*(\varphi*,k)} \Pi_{t*}^*$$
$$+ (\nabla_{u*} b^{*(\varphi*,k)})_o \Pi^* + b_{ox}^{*(\varphi*,k)} X^* = 0, \tag{5.38}$$

for $k = 1,2,\ldots,m_1^*$, and

$$\frac{dX^*}{dt^*} = (\nabla_{u*}\lambda^{*(\varphi*)})_o \Pi^* + \lambda_{ox}^{*(\varphi*)} X^*. \tag{5.39}$$

When initial conditions are given in $\mathscr{R}_u$ and $\mathscr{R}_{u*}$ then sufficient information is available to determine $\Pi(t')$ and $\Pi^*(t^*)$. These discontinuity vectors give the required information about the variation of the propagating C^1 discontinuities in U and U^*. To complete the analysis in $\mathscr{R}_u$ and $\mathscr{R}_{u*}$ it only remains for us to relate the vectors Π and Π^*.

5.5 Conditions Across The Shock Line $\mathscr{D}$

We presuppose that the solutions $U_o(x,t)$ and $U_o^*(x,t)$ are known together with the shock line $\mathscr{D}$, and that systems (5.1) and (5.2) are of conservation type so that across $\mathscr{D}$ the generalized jump condition (5.4) is valid. This condition, which through F, F^* relates the undifferentiated vectors U, U^* and the vectors G, G^*, which are nonlinear functions of U and U^*, may be used to relate the gradients of U and U^* across $\mathscr{D}$ as we now show. Henceforth we assume that the uniqueness of $U_o^*(x,t)$ has been assured by appeal to the evolutionary condition when selecting the smooth piecewise continuous weak solution appropriate to (5.4).

Before deriving the relationship between the gradients of U and U^* across $\mathscr{D}$, it is first necessary that we consider the behaviour of the characteristics to the left and right of $\mathscr{D}$ at P. Immediately to the left of $\mathscr{D}$ at P there will, in general, be q characteristics belonging to system (5.1) which, as time increases, radiate out from P and enter $\mathscr{R}_u$ to the left of $\mathscr{D}$; the remaining p-q characteristics belonging to system (5.1) will lie to the right of $\mathscr{D}$ and so would, if produced, enter $\mathscr{R}_{u*}$. Analogously, immediately to the right of $\mathscr{D}$ at P there will, in general, be q* characteristics belonging to system (5.2) which, as time increases, radiate out from P and enter $\mathscr{R}_{u*}$ to the right of $\mathscr{D}$; the remaining p^*-q^* characteristics belonging to system

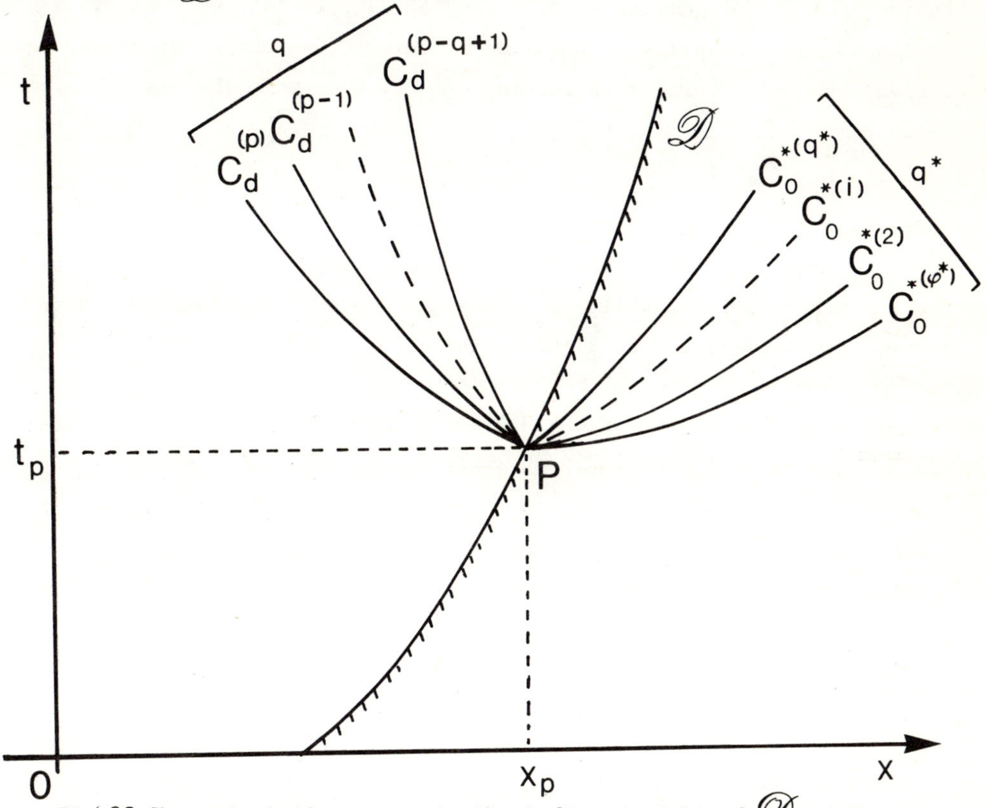

Fig. 23 Characteristic curves to the left and right of $\mathscr{D}$ at P.

(5.2) will lie to the left of $\mathscr{D}$ and so would, if produced, enter $\mathscr{R}_{u*}$. Figure 23 shows only those members of the family C of characteristic curves that lie to the left of $\mathscr{D}$ at P, and those of the family C^* of characteristic curves that lie to the right of $\mathscr{D}$ at P.

It follows from this that since C^1 discontinuities can only be transmitted along characteristics, then in principle C^1 discontinuities may be transmitted forwards into $\mathscr{R}_{u*}$ along the q^* characteristics $C_o^{*(\varphi*)}$, $C_o^{*(2)}, \ldots C_o^{*(q*)}$, and reflected back into $\mathscr{R}_u$ along the q characteristics $C_d^{(p-q+1)}$, $C_d^{(p-q+2)}, \ldots, C_d^{(p)}$.

No other possibilities exist, and the actual situation that results in any particular case must be resolved by examination of the manner in which C^1 discontinuities are propagated in the regions adjacent to $\mathscr{D}$.

To the left of $\mathscr{D}$ at P, let the reflected C^1 discontinuity vectors transported along the characteristics $C_d^{(p-q+i)}$, with i = 1,2,...,q be denoted by $\Pi_i^{(R)}$. Then as the incident C^1 discontinuity vector transported along $C_o^{(\varphi)}$ is the vector Π of equations (5.32) and (5.34), it follows that since jumps in the normal derivative of U may occur only across characteristics, then across the q + 1 characteristics in $\mathscr{R}_u$ at P we have the result

$$\Pi(P) + \sum_{i=1}^{q} \Pi_i^{(R)}(P) = U_\varphi^{(R)}(P) - U_{\varphi o}(P). \tag{5.40}$$

Here $U_{\varphi o}$ signifies the value of U_φ in $\mathscr{R}_u$ just ahead of $C_o^{(\varphi)}$ appropriate to the known solution U_o, whilst $U_\varphi^{(R)}$ is the value of the related vector U_φ in $\mathscr{R}_u$ immediately to the right of the characteristic curve $C_d^{(p-q+1)}$ in Fig. 23 .

Analogously, to the right of $\mathscr{D}$ at P, let the C^1 discontinuity vectors transported along the characteristics $C_o^{*(i)}$ be denoted by Π_i^*, with the understanding that $\Pi_1^* \equiv \Pi^*$, the vector occurring in equations (5.37) and (5.38). Then across the q^* characteristics in $\mathscr{R}_{u*}$ at P we also have

$$\sum_{i=1}^{q*} \Pi_i^*(P) = U_{\varphi*}^{*(T)}(P) - U_{\varphi o*}^*(P). \tag{5.41}$$

Here $U_{\varphi o*}^*$ signifies the value of $U_{\varphi*}^*$ in $\mathscr{R}_{u*}$ just ahead of $C_o^{*(\varphi*)}$ appropriate to the known solution U_o^*, whilst $U_{\varphi*}^{*(T)}$ is the value of the transmitted vector $U_{\varphi*}^*$, in $\mathscr{R}_{u*}$ to the left of the characteristic $C_j^{*(q*)}$.

Two tasks now remain, the first being that of relating $U_\varphi^{(R)}(P)$ and $U_{\varphi*}^{*(T)}(P)$ across $\mathscr{D}$, and the second that of determining the vectors $\Pi_i^{(R)}$ and Π^*.

The first of these tasks will be accomplished by appeal to the jump condition (5.4), which we differentiate with respect to t at P. After using the fact that $dx/dt = \tilde{\lambda}$ along $\mathscr{D}$, we obtain the equation

$$\{G_t + \tilde{\lambda}G_x + (\nabla_u G)(U_t + \tilde{\lambda}U_x)\}_p - \{G^*_t + \tilde{\lambda}G^*_x + (\nabla_{u*} G^*)(U^*_t + \tilde{\lambda}U^*_x)\}_p$$

$$= \left\{\frac{d\tilde{\lambda}}{dt}\right\}_p \{F - F^*\}_p + \tilde{\lambda}(P)\{U_t + \tilde{\lambda}U_x + (\nabla_u F)(U_t + \tilde{\lambda}U_x)\}_p \qquad (5.42)$$

$$- \tilde{\lambda}(P)\{U^*_t + \tilde{\lambda}U^*_x + (\nabla_{u*}F^*)(U^*_t + \tilde{\lambda}U^*_x)\}_p.$$

Here G_t signifies partial differentiation of G with respect to t with $x, u_1, u_2, \ldots, u_n$ regarded as constants, and G_x and the corresponding starred derivatives are similarly defined.

Using (5.1) and (5.2) together with the fact that to the left of $\mathscr{D}$ at P, $U_x(P) = U_x^{(R)}(P)$, whilst to the right $U^*_x(P) = U_x^{*(T)}(P)$, we obtain the required relationship in the form

$$\left\{G_t + \tilde{\lambda}G_x - \left\{\frac{d\tilde{\lambda}}{dt}\right\}_p F + (\tilde{\lambda}\nabla_u F - \nabla_u G)[(A - \tilde{\lambda}I)U_x^{(R)} + B]\right\}_p$$

$$= \left\{G^*_t + \tilde{\lambda}G^*_x - \left\{\frac{d\tilde{\lambda}}{dt}\right\}_p F^* + (\tilde{\lambda}\nabla_{u*}F^* - \nabla_{u*}G^*)[(A^* - \tilde{\lambda}I)U_x^{*(T)} + B^*]\right\}_p. \qquad (5.43)$$

The two derivatives $U_\varphi^{(R)}(P)$ and $U_{\varphi*}^{*(T)}(P)$ are then related via (5.43) by using the fact that

$$U_x^{(R)}(P) = \frac{U_\varphi^{(R)}(P)}{x_\varphi(P)} \quad \text{and} \quad U_x^{*(T)}(P) = \frac{U_{\varphi*}^{*(T)}(P)}{x_{\varphi*}(P)}. \qquad (5.44)$$

On account of these results we may re-express equations (5.40) and (5.41) in the form

$$\frac{\Pi(P)}{x_\varphi(t_p)} + \frac{1}{x_\varphi(t_p)}\sum_{i=1}^{q}\Pi_i^{(R)}(P) = U_x^{(R)}(P) - U_{ox}(P)\left[\frac{x_{\varphi o}(t_p)}{x_\varphi(t_p)}\right] \qquad (5.45)$$

or, from the definition of Π,

$$U_x^{(R)}(P) = U_x(P) + \frac{1}{x_\varphi(t_p)}\sum_{i=1}^{q}\Pi_i^{(R)}(P),$$

and

$$\frac{1}{x_{\varphi*}(P)}\sum_{i=1}^{q^*}\Pi^*_i(P) = U_x^{*(T)}(P) - U^*_{ox}(P)\left[\frac{x_{\varphi o*}(P)}{x_{\varphi*}(P)}\right]. \qquad (5.46)$$

From the initial condition (5.23) we see that $x_{\varphi o*}(P) = x_{\varphi*}(P) = 1$, and it then follows both that $X^*(P) = 0$ and that (5.46) simplifies to give

$$\frac{1}{x_{\varphi*}(P)} \sum_{i=1}^{q*} \Pi_i^*(P) = U_x^{*(T)}(P) - U_{ox}^*(P). \tag{5.47}$$

In this result the unit factor $1/x_{\varphi*}(P)$ has been retained on the left hand side to clarify the relationship between derivatives with respect to x and φ^*. Taken together, equations (5.43), (5.45) and (5.47) determine the relationship between $U_x^{(R)}(P)$ and $U_x^{*(T)}(P)$ once the vectors $\Pi_i^{(R)}$ and Π_i^* are known.

Now to the right of $\mathscr{D}$ at P the orthogonality of the left and right eigenvectors of the matrix A_o^*, taken together with (5.37) which is valid along $C_o^{*(\varphi*)}$ in $\mathscr{R}_{u*}$, implies that $\Pi^*(t^*) - U_{ox}^*(t^*)X^*(t^*)$ is linearly dependent on the m_1^* eigenvectors $r_o^{*(\varphi*,k)}(t^*)$, $k = 1,2,\ldots,m_1^*$ corresponding to the eigenvalue $\lambda^{*(\varphi*)}$. Thus we may write

$$\Pi^*(t^*) - U_{ox}^*(t^*)\, X^*(t^*) = \sum_{k=1}^{m*_1} \alpha_k^{(\varphi*)} r_o^{*(\varphi*,k)}(t^*), \tag{5.48}$$

where the $\alpha_k^{(\varphi*)}$ are constants, and

$$A_o^*(t^*)\; r_o^{*(\varphi*,k)}(t^*) = \lambda_o^{*(\varphi*)} r_o^{*(\varphi*,k)}(t^*) \tag{5.49}$$

for $k = 1,2,\ldots,m_1^*$.

Correspondingly, along any other one of the q^* characteristics along which propagation in $\mathscr{R}_{u*}$ is possible, say $C_o^{*(i)}$ occurring with multiplicity m_i^*, the expression analogous to (5.48) is obtained by replacing φ^* by i and m_1^* by m_i^* to yield

$$\Pi_i^*(t^*) - U_{ox}^*(t^*)\, X^*(t^*) = \sum_{k=1}^{m_i^*} \alpha_k^{(i)} r_o^{*(i,k)}(t^*), \tag{5.50}$$

for $i = 1,2,\ldots,q^*$. As before, the understanding here is that the superscript φ^* in (5.48) corresponds to the case $i = 1$.

The eigenvector $r_o^{*(i,k)}(t^*)$ satisfies an equation derived from (5.49) by replacing φ^* by i and m_1^* by m_i^*. It has already been remarked that $X^*(P) = 0$, so that by setting $t^* = t_p$, (5.50) simplifes to give

$$\Pi_i^*(P) = \sum_{k=1}^{m_i^*} \alpha_k^{(i)}\, r_o^{*(i,k)}(P). \tag{5.51}$$

A corresponding argument applied to the left of $\mathscr{D}$ at P gives the similar result

$$\Pi_i^{(R)}(P) = \sum_{k=1}^{m_{p-q+i}} \beta_k^{(i)}\, r_o^{(i,k)}(P), \tag{5.52}$$

where P has been taken as the origin for the reflected waves, the characteristics of which are parameterised in such a manner that the scalar jump quantity $X^{(R)}$ analogous to X^* has the property that $X^{(R)}(P) = 0$, with $X_\varphi^{(R)}(P) = x_\varphi(t_p)$.

Equations (5.43), (5.45), (5.47), (5.51) and (5.33) imply n inhomogeneous algebraic equations for the N constants $\alpha_1^{(1)}, \alpha_2^{(1)}, \ldots, \alpha_{m_q*}^{(q*)}, \beta_1^{(1)}, \beta_2^{(1)}, \ldots, \beta_{m_q}^{(q)}$, where

$$N = \sum_{i=1}^{q*} m_i^* + \sum_{j=1}^{q} m_j. \tag{5.53}$$

If the rank of the coefficient matrix (5.43) is $\tilde{n}$, then a solution comprising a set of N such constants will only exist, and be determined in a unique non-trivial manner, if $N = \tilde{n}$. This will be called the normal case, and it will be the only one considered here. In passing, it is of interest to note that this condition is, in some ways, the analogue of the evolutionary condition. Once the coefficients $\alpha_k^{(i)}, \beta_k^{(i)}$ have been determined, equations (5.51) then give the initial values $\Pi_i^*(P)$ of the transmitted C^1 discontinuity vectors $\Pi_i^*(t^*)$. In general, only propagation along $C_o^{*(\varphi*)}$ will be the object of study, so that usually the initial vector $\Pi_1^*(P)$ will be the only one that is required. The behaviour of the reflected C^1 discontinuities would be difficult to examine, since in order to do so it would be necessary to know the solution in $\mathscr{R}_u$ after the incident C^1 discontinuity associated with the finite amplitude wave had passed.

5.6 Formation of Shock on the Wavefront

We begin by supposing that the C^1 discontinuity on the initial line $t = 0$ has propagated to point P on $\mathscr{D}$ without the Jacobian x_φ vanishing, and hence that propagation along $C_o^{*(\varphi*)}$ in $\mathscr{R}_{u*}$ is to be considered. The linear ordinary differential equation (5.36) is thus valid everywhere along $C_o^{(\varphi)}$ in $\mathscr{R}_u$ up to point P. The integrating factor μ for this equation is

$$\mu(t') = \exp\{-\int \lambda_{ox}^{(\varphi)} dt'\},$$

so that

$$\frac{d}{dt'} (\mu X) = \mu(\nabla_u \lambda^{(\varphi)})_o \Pi$$

or, integrating from 0 to t,

$$\mu(t)\, X(t) = \mu(0)\, X(0) + \int_o^t \mu(t')\, (\nabla_u \lambda^{(\varphi)})_o\, \Pi(t')dt'. \tag{5.54}$$

However, from the initial condition (5.20), on the initial line $x_\varphi\big|_{\varphi=0_-} = x_\varphi\big|_{\varphi=0_+} = 1$, so that $X(0) = 0$ so (5.54) reduces to

$$X(t) = \frac{1}{\mu(t)} \int_o^t \mu(t')\, (\nabla_u \lambda^{(\varphi)})_o \Pi(t')dt', \tag{5.55}$$

thereby giving a direct expression for X(t).

An equation analogous to (5.54) can be derived along $C_o^{*(\varphi*)}$ in $\mathscr{R}_{u*}$ by integration of (5.39) from t_p to τ, when we find

$$\mu^*(\tau)\, X^*(\tau) = \mu^*(t_p)\, X^*(t_p) + \int_{t_p}^{\tau} \mu^*(t')(\nabla_{u*}\lambda^{*(\varphi*)})_o \Pi^*(t')dt'. \tag{5.56}$$

However, as we have already observed in Section 5.5, from the initial condition (5.23) assumed for φ^* along $t = t_p$ it follows that $x_{\varphi*}(t_p) = x_{\varphi o*}(t_p) = 1$, so that $X^*(t_p) = 0$.

Now, by definition, $X^*(\tau) = x_{\varphi*}(\tau)\big|_{\varphi*=0_-} - x_{\varphi*}(\tau)\big|_{\varphi*=0_+} = x_{\varphi*}(\tau)\big|_{\varphi=0_-} - x_{\varphi o*}(\tau)$. So if we now write

$$x_{\varphi o*}(\tau) = \tilde{x}_{\varphi*} h^*(\tau) \quad \text{where } \tilde{x}_{\varphi*} = x_{\varphi o*}(t_p),$$

then after division by $\mu^*(\tau)$ and use of these results equation (5.56) takes the form

$$x_{\varphi*}(\tau)\big|_{\varphi*=0_-} = \tilde{x}_{\varphi*} h^*(\tau) + \frac{1}{\mu^*(\tau)} \int_{t_p}^{\tau} \mu^*(t')(\nabla_{u*}\lambda^{*(\varphi*)})_o \Pi^*(t')dt'. \tag{5.57}$$

The left hand side of (5.57) is simply the Jacobian of the coordinate transformation used in $\mathscr{R}_{u*}$, so that if for some time $\tau = t_c$ this Jacobian vanishes, the $C^{*(\varphi*)}$ family of characteristic curves must intersect on the wavefront trace and a singularity then occurs in the transformation, and possibly in the solution as well. We postpone comment on the nature of this singularity until the end of this section. Hence, if such a time t_c exists, it is determined by the equation

$$0 = \tilde{x}_{\varphi*}(t_c) + \frac{1}{\mu^*(t_c)} \int_{t_p}^{t_c} \mu^*(t')(\nabla_{u*}\lambda^{*(\varphi*)})_o \Pi^*(t')dt'. \tag{5.58}$$

In the special case that the coefficients of (5.1) and (5.2) are both continuous and continuously differentiable across $\mathscr{D}$, so that no discontinuity exists, then equation (5.55) with $t = t_p$ and (5.58) combine to yield

$$0 = \tilde{x}_\varphi h(t_c) + \frac{1}{\mu(t_c)} \int_o^{t_c} \mu(t')(\nabla_u\lambda^{(\varphi)})_o \Pi(t')dt', \tag{5.59}$$

where $x_{\varphi o}(t) = \tilde{x}_\varphi h(t)$, with $\tilde{x}_\varphi = x_{\varphi o}(0)$. Equations (5.32), (5.34), and (5.59) reduce to those given in earlier work [8, 116] when $U_o(x,t) = U_o$, a constant, or when $\lambda_{ox}^{(\varphi)} \equiv 0$ and $\ell_{ox}^{(\varphi,k)} = b_{ox}^{(\varphi,k)} \equiv 0$, $k = 1,2,\ldots,m_1$. In both of these cases $\mu \equiv 1$. When $U_o(x,t) = U_o$, a constant, a further simplification occurs for then $h \equiv 1$.

It remains for us to deduce the form of the function $h^*(\tau)$ occurring in (5.57) which is, in effect, a scale factor relating the old and new independent variables x, φ^* in $\mathscr{R}_{u*}$ ahead of $C_o^{*(\varphi*)}$. From (5.15) the C^* characteristic through the point (ξ, t_p) is determined by

$$t^* - t_p = \int_\xi^x \frac{ds}{\lambda_o^{*(\varphi*)}(s)},$$

where the suffix O refers to the solution U_o^* immediately ahead of this C^* characteristic. As x, ξ are functions of φ^*, differentiating this expression with respect to φ^* and using the result $\partial t^*/\partial\varphi^* \equiv 0$ gives

$$0 = \frac{\partial x}{\partial\varphi^*}\left[\frac{1}{\lambda_o^{*(\varphi*)}(x)}\right] - \frac{\partial\xi}{\partial\varphi^*}\left[\frac{1}{\lambda_o^{*(\varphi*)}(\xi)}\right].$$

However, the parameterisation of φ^* in (5.23) assigns to the C^* characteristic through (ξ, t_p) the value $\varphi^* = \xi - x_p$, so that $\partial\xi/\partial\varphi^* = 1$. Consequently on the wavefront trace $C_o^{*(\varphi*)}$ corresponding to $\xi = x_p$ this expression becomes

$$x_{\varphi o*}(x) = \lambda_o^{*(\varphi*)}(x)/\lambda_o^{*(\varphi*)}(x_p) \text{ along } C_o^{*(\varphi*)} \tag{5.60}$$

or, equivalently in terms of t^*,

$$x_{\varphi o*}(t^*) = \lambda_o^{*(\varphi*)}(t^*)/\lambda_o^{*(\varphi*)}(t_p) \text{ along } C_o^{*(\varphi*)}. \tag{5.61}$$

As we have set

$$x_{\varphi o*}(\tau) = \tilde{x}_{\varphi*}h^*(\tau)$$

it then follows from (5.61) that

$$\tilde{x}_{\varphi*}h^*(\tau) = \lambda_o^{*(\varphi*)}(\tau)/\lambda_o^{*(\varphi*)}(t_p) \text{ along } C_o^{*(\varphi*)}. \tag{5.62}$$

Although initial condition (5.23) implies $\tilde{x}_{\varphi*} = 1$, as before, to clarify the relationship between $U^*_{\varphi*}$ and U^*_x, it is convenient to retain the symbol $\tilde{x}_{\varphi*}$. An alternative derivation of result (5.26) could have been based on the direct integration of the first equation of (5.22), subject to initial condition (5.23). The scale factor $h(\tau)$ is similarly defined.

The x-coordinate x_c of the singularity on the wavefront trace at time $t = t_c$ is given by

$$x_c = x_p + f(x_p, t_c), \tag{5.63}$$

where $x = f(x_p, t)$ is simply the equation of the wavefront trace $C_o^{*(\varphi*)}$ through P expressed as a function of t.

The nature of the singularity occurring at (x_c, t_c) is, in most applications, usually the formation of a shock in U^*. This will be the case if $U^*_{\varphi*}$ is finite at $t = t_c$ but $x_{\varphi*} = 0$, for then, just behind the wavefront trace, $U^*_x = U^*_{\varphi*}/x_{\varphi*}$ becomes infinite, and differentiability breaks down. A shock can also arise from $U^*_{\varphi*}$ itself becoming infinite either before $x_{\varphi*}$ vanishes or simultaneously with it. The former occurrence must be investigated separately, by examining the behaviour of $U^*_{\varphi*}$ once Π^* is known, whilst the latter is covered by examining the vanishing of $x_{\varphi*}$. It may happen that when the Jacobian $x_{\varphi*}(t_c)\big|_{\varphi*=0_-} = 0$, it is also the case that $U^*_{\varphi*} = 0$. The nature of the singularity at t_c will then be determined by the limit of the indeterminate form $U^*_{\varphi*}/x_{\varphi*}$ as $t \to t_c$. If this is infinite, then it corresponds to the formation of a shock in U^*, whereas if it is finite no shock forms and it corresponds only to a singularity of the transformation used. Should it occur, the anomaly may be removed by adopting a different parameterization for φ^* in the initial condition (5.23).

5.7 Special Cases

There are a number of special cases of the foregoing analysis which are worthy of note, and we now outline the simplest of these. As will be seen, some arise as a result of the special structure of equations (5.1) and (5.2) as determined by particular forms of matrices A, B, A^* and B^*, whilst others are of an entirely new nature and occur because of the presence of the shock line $\mathscr{D}$.

(i) <u>Homogeneous Case $B = B^* = 0$</u>

Some simplification in the manipulation results when equations (5.1) and (5.2) are homogeneous, so that $B = B^* \equiv 0$. It then follows directly that

$$U_o(x,t) = U_o \text{ and } U_o^*(x,t) = U_o^*, \tag{5.64}$$

where U_o, U_o^* are constant vectors. We deal first with the implications of this in $\mathscr{R}_u$. Equations (5.32) and (5.34) reduce, respectively, to

$$\ell_o^{(i,k)}\Pi = 0, \tag{5.65}$$

for $i = 2,3,\ldots,p$ and $k = 1,2,\ldots,m_i$, and

$$\ell_o^{(\varphi,k)}\Pi_{t'} = 0, \tag{5.66}$$

for $k = 1,2,\ldots,m_1$.

As the $\ell_o^{(i,k)}$ are constant vectors, differentiation of (5.65) with respect to t' gives

$$\ell_o^{(i,k)}\Pi_{t'} = 0, \tag{5.67}$$

for $i = 2,3,\ldots,p$ and $k = 1,2,\ldots,m_i$. Now as, by hypothesis, there is a full set of the left eigenvectors of A_o, and they are linearly independent, the n homogeneous algebraic equations (5.66) and (5.67) can only have the solution $\Pi_{t'} = 0$, showing that Π is a constant equal to its initial value

$$\Pi = \tilde{\Pi}. \tag{5.68}$$

An exactly similar argument shows that in $\mathscr{R}_{u*}$ the vector Π^* is a constant and equal to its initial value

$$\Pi^* = \tilde{\Pi}^*. \tag{5.69}$$

As $U_{\varphi o} \equiv 0$ it follows that $\Pi = U_\varphi$, and so by virtue of (5.68) we have $\Pi = \tilde{U}_\varphi$, a constant, with the corresponding result $\Pi^* = \tilde{U}^*_{\varphi*}$ in $\mathscr{R}_{u*}$. Thus we find that

$$U_x = \tilde{U}_\varphi / x_\varphi \text{ and, similarly, } U_x^* = U_{\varphi*}^* / x_{\varphi*}. \quad (5.70)$$

Once it has been established from equation (5.55) that no forms before is reached, equations (5.43), (5.45), (5.47), (5.51) and (5.52) may be solved to determine $\Pi_i^*(P) \equiv \Pi^*(P)$. The subsequent analysis leading to the determination of t_c and x_c on $C^{*(\varphi*)}$ then proceeds as before, though the task is now simplified by the fact that $h^* \equiv 1$.

(ii) Coefficient Matrices Not Explicitly Dependent on x and t

If, in $\mathscr{R}_u$, the coefficient matrices, A, B are not explicitly dependent on x and t, it follows from (5.1) that in the region ahead of the wavefront trace $C_o^{(\varphi)}$, $U(x,t) = U_o$ where U_o is a constant vector. The only restriction on U_o is that

$$B(U_o) = 0. \quad (5.71)$$

These results imply that the vectors $U_{ot'} = U_{ox} = b_{ox} \equiv 0$ and that $\lambda_{ox} \equiv 0$, $\mu \equiv 1$ and $h \equiv 1$, so that equations (5.32), (5.34) and (5.55) simplify, respectively, to

$$\ell_o^{(i,k)} \Pi = 0 \quad (5.72)$$

for $i = 2,3,\ldots,p$ and $k = 1,2,\ldots,m_i$,

$$\ell_o^{(\varphi,k)} \Pi_{t'} + (\nabla_u b^{(\varphi,k)})_o \Pi = 0, \quad (5.73)$$

for $k = 1,2,\ldots,m_1$, and

$$X(t) = \int_o^t (\nabla_o \lambda^{(\varphi)})_o \Pi(t')dt'. \quad (5.74)$$

If, in $\mathscr{R}_{u*}$, the coefficient matrices A^*, B^* are not explicitly dependent on x and t, then similar results apply there, though in place of (5.58), since $h^* \equiv 1$, we now have

$$0 = \tilde{x}_{\varphi*} + \int_{t_p}^{t_c} (\nabla_{u*} \lambda^{*(\varphi*)})_o \Pi^*(t')dt'. \quad (5.75)$$

Once the initial value $\Pi^*(P)$ of the vector $\Pi^*(t')$ occurring in equation (5.75) has been found in the manner already described, then the vector $\Pi^*(t')$ itself may be obtained by solving the starred equations analogous to (5.72) and (5.73) and again using results (5.70) which are still valid in this case. As $h \equiv 1$

the quantity $x_\varphi(t_p)$ needed in (5.45) follows directly from (5.14) in the form

$$x_\varphi(t_p) = \tilde{x}_\varphi + \int_0^{t_p} (\nabla_u \lambda^{(\varphi)})_0 \Pi(t')dt', \tag{5.76}$$

where $\tilde{x}_\varphi = x_{\varphi 0} \equiv 1$, though we retain the symbol $\tilde{x}_\varphi$ here to make clear the relationship between x and φ.

(iii) Non-occurrence of Shock - Exceptional System

It can happen that although a C^1 discontinuity propagates, no shock forms on the wavefront trace in $\mathscr{R}_{u*}$ at any time t_c. Such an exceptional situation may come about for a number of reasons of which we choose to mention only the most straightforward useful cases. We shall assume that $U^*_{\varphi*}$ remains finite and that the occurrence of a shock in U^* corresponds to the vanishing of $x_{\varphi*}$. This assumption implies no restriction, because it can be verified by determining Π^* and hence $U^*_{\varphi*}$. In the first instance we mention, the time t_c at which the Jacobian vanishes may be a function of one or more parameters characterising the problem. When this occurs it is possible that for some parameter values the time t_c will become infinite. In physical problems this is likely to correspond to a competition between two different processes, one leading to the steepening of a wave and the other to its relaxation. Such is the case, for example, when water waves propagate into deepening water, for then the nonlinearity in the equations tends to produce breaking of the waves, whilst the deepening produces the relaxation mentioned. For any given wave a certain gradient m of sea-bed exists, which, when attained or exceeded, will prevent the breaking of the water wave in question. A study of this problem has been made by Jeffrey [133]. In gas dynamics we cite as an example of this competitive process the interaction between the normal steepening effect in a compression wave and the dissipative effects of relaxation in a gas [140].

In contrast to these inherently nonlinear systems in which the formation of a strong discontinuity is normal, but may be averted in particular cases by the choice of special parameter values characterising the problem, there is an important and easily identifiable class of problems in which shocks can never occur on the wavefront trace. These are the situations in which $(\nabla_{u*}\lambda^{*(\varphi*)})_0$ and $\Pi^*(t')$ are orthogonal, so that $(\nabla_{u*}\lambda^{*(\varphi*)})_0\Pi^*(t') \equiv 0$, and the condition

$$h^*(\tau) \neq 0 \tag{5.77}$$

is also true for $\tau > t_p$. This latter condition results from (5.57) by setting the integral identically equal to zero, and then requiring that

$$x_{\varphi^*}(\tau)\Big|_{\varphi^*=0_-} \neq 0 \quad \text{for } \tau > t_p.$$

Now in the completely general case involving no assumptions about A^*, B^*, equation (5.48) is still true, so that the vector $\Pi^* - U^*_{ox}X^*$ at any time $t^* > t_p$ is a linear combination of the m^*_1 right eigenvectors $r_o^{*(\varphi^*,k)}$, k = 1,2, ...,m^*_1 at that time t^*. Hence, if $(\nabla_{u^*}\lambda^{*(\varphi)}_1)_o$ is orthogonal to both the $r^{*(\varphi^*,k)}$, $k = 1,2,\ldots,m^*_1$ and to U^*_{ox} for all time, it follows that it must be orthogonal to Π^* for all time, and the conditions leading to the non-existence of a time t_c will be realised.

Of the many ways in which this orthogonality can arise, probably the simplest is when the first s components of U^*_{ox} and of the m^*_1 vectors $r^{*(\varphi,k)}$ are zero, and $\lambda^{*(\varphi^*)}$ is independent of the last (n - s) elements u^*_i of vector U^*, for then the last (n - s) elements of $(\nabla_{u^*}\lambda^{*(\varphi^*)})_o$ will also be zero.

This special structure corresponds to a matrix A^* which is composed of invariant subspaces of dimension s and n - s, respectively, so that in partioned form it becomes

$$A^* = \begin{bmatrix} A^*_1 & \vdots & 0 \\ \cdots\cdots & \vdots & \cdots\cdots \\ 0 & \vdots & A^*_2 \end{bmatrix} \qquad (5.78)$$

where matrix A^*_1 is square and of order s, and A^*_2 is square and of order (n - s). The n characteristic roots of A^* comprise the s characteristic roots $\lambda^{*(q)}_1$ of A^*_1 and the (n - s) characteristic roots $\lambda^{*(q)}_2$ of A^*_2.

In the general case that the eigenvalue $\lambda = \lambda^{*(i)}$ of A^*_2 occurs with multiplicity m^*_{2i}, and has the associated eigenvectors $r^{*(i,k)}$, $k = 1,2,\ldots,m^*_{2i}$, we know that in partitioned form

$$r^{*(i,k)} = \begin{bmatrix} 0 \\ \vdots \\ 0 \\ r_2^{*(i,k)} \end{bmatrix} \quad \text{for } k = 1,2,\ldots,m^*_{2i} \qquad (5.79)$$

where $r_2^{*(i,k)}$ is the right ((n - s) $\times$ 1) eigenvector of A^*_2 corresponding to $\lambda = \lambda_2^{*(i)}$.

If A^*_2 is independent of the last (n - s) elements u^*_i of U^*, then so also will be the eigenvalue $\lambda_2^{*(i)}$, and $(\nabla_{u^*}\lambda_2^{*(i)})_o$ will have zero elements in its last (n - s) positions. Identifying $\lambda^{*(i)}$ with $\lambda^{*(\varphi*)}$ and m^*_{2i} with m^*_1 we arrive at the required structure for $(\nabla_{u*}\lambda^{*(\varphi*)})_o$.

Whenever

$$(\nabla_{u*}\lambda^{*(i)})_o r_o^{*(i,k)} = 0, \text{ for } k = 1,2,\dots,m^*_{2i}, \tag{5.80}$$

then using the terminology of Definition 3.4 we shall say that system (5.2) is exceptional with respect to the $\lambda^{*(i)}$ field of characteristics. Thus systems of equations (5.1) and (5.2) will not generate a shock on $C_o^{*(\varphi*)}$ in $\mathscr{R}_{u*}$ if system (5.2) is exceptional with respect to the $\lambda^{*(\varphi*)}$ field of characteristics, and the first s elements of both $r_o^{*(\varphi*,k)}$, $k = 1,2,\dots,m^*_1$ and U^*_{ox} are zero.

Typical examples of exceptional systems with respect to one particular eigenvalue, in which A^* has the structure just mentioned, are the contact discontinuity in adiabatic gas flow and the entropy and Alfven waves in magnetohydrodynamics [8].

An interesting new physical example which does not involve discontinuous coefficients has been provided by Rhee, Aris and Amundson [55] in connection with the mathematical analysis of multicomponent chromatography. The equations involved are homogeneous, thereby permitting the introduction of generalised Riemann invariants, so that discontinuities of the contact type are also possible in principle. These have been identified and examined explicitly by Rhee and Amundson in another paper in connection with an adiabatic adsorption column [122].

(iv) Shock $\mathscr{D}$ is $\lambda^{*(i)}$ Wave Suppressive

The phenomena described in this and the final sub-section are essentially new, and occur only because of the existence of the shock line $\mathscr{D}$ in the (x,t)-plane. In Section 5.5 we established how a C^1 discontinuity transmitted forwards into $\mathscr{R}_{u*}$ along the characteristic $C_o^{*(i)}$, one of the q* characteristics $C_o^{*(\varphi*)}, C_o^{*(2)}, \dots, C_o^{*(q*)}$ entering $\mathscr{R}_{u*}$ from P, does so with an initial C^1 discontinuity vector $\Pi^*_i(P)$. Normally, propagating C^1 discontinuities will be initiated along each of the q* characteristics in question, but if for some characteristic $C_o^{*(i)}$ it happens that $\alpha_1^{(i)} = \alpha_2^{(i)} = \dots = \alpha_{m_i*}^{(i)} = 0$ then no C^1 discontinuity can be propagated along the characteristic. When this happens we shall say that the shock line $\mathscr{D}$ is $\lambda^{*(i)}$ wave suppressive. In the exceptional case that this is true for each of the q* characterisitcs entering

$\mathscr{R}_{u*}$ from P, we shall say that the line $\mathscr{D}$ is completely λ^* wave suppressive so that no C^1 discontinuity is then propagated across $\mathscr{D}$ into $\mathscr{R}_{u*}$. Alternatively, the line $\mathscr{D}$ may be termed totally λ wave reflective. A simple example of this occurs when an elastic wave propagating in a layered medium encounters a rigid wall which then acts as a reflecting barrier. A very general problem of this type has been investigated by Jeffrey and Suhubi [120] with special reference to the transmission and reflection properties at the successive interfaces. The rigid boundary just mentioned follows as a trivial special case of this analysis.

(v) <u>Shock $\mathscr{D}$ is $\lambda^{(i)}$ Wave Non-Reflective</u>

Here we consider the situation in $\mathscr{R}_u$ at P after the initiation of the q reflected waves along the characteristics $C_d^{(p-q+1)}$, $C_d^{(p-q+2)}, \ldots, C_d^{(p)}$ that will normally be associated with the arrival of the incident C^1 discontinuity along $C_o^{(\varphi)}$. Again appealing to the results of Section 5.5, we can find the initial value $\Pi_i^{(R)}(P)$ of the reflected C^1 discontinuity vector $\Pi_i^{(R)}$ propagated along the characteristic $C_d^{(p-q+i)}$.

If, for some i, it happens that $\beta_1^{(i)} = \beta_2^{(i)} = \ldots = \beta_{mi}^{(i)} = 0$ then no C^1 discontinuity can be reflected back along that characteristic. When this happens we shall say that the shock line $\mathscr{D}$ is $\lambda^{(p-q+i)}$ wave non-reflective. In the exceptional case that this is true for each of the q characteristics entering $\mathscr{R}_u$ from P, we shall say that the shock line $\mathscr{D}$ is completely λ wave non-reflective. Alternatively, and more positively, the line $\mathscr{D}$ may be termed totally λ wave transmissive.

5.8 <u>Bifurcation of Wavefront</u>

In formulating the problem in Section 5.1 we required that the multiplicities m_i and m_i^*, of the p distinct eigenvalues $\lambda^{(i)}$ and the p^* distinct eigenvalues $\lambda^{*(i)}$, be constant in the regions $\mathscr{R}_u$ and $\mathscr{R}_{u*}$, respectively. As the matrices A, A^* are not constant coefficient matrices it may happen that this assumption is untrue for some systems. Let us suppose that these multiplicities change at some point on the wavefront trace, then the propagating C^1 discontinuity will bifurcate and new C^1 discontinuities will propagate. This situation is in some ways similar to the one described in Section 5.5, and it is to that section that we must turn for the method which will enable us to lift the restriction to constant multiplicities. As bifurcation may happen either on $C_o^{(\varphi)}$ or on $C_o^{*(\varphi*)}$, and the problem involved is identical in each case, we shall consider only bifurcation on $C_o^{(\varphi)}$ in $\mathscr{R}_u$.

Consider system (5.1) in region $\mathscr{R}_u$, and assume that the C^1 discontinuity propagating along $C_o^{(\varphi)}$ bifurcates in the postulated manner at the point Q with coordinates (x_q, t_q). Suppose, further, that the p distinct eigenvalues $\lambda^{(1)}, \lambda^{(2)}, \ldots, \lambda^{(p)}$ of A occurring with multiplicities $m_1, m_2, \ldots, m_p$, of which $\lambda^{(1)} \equiv \lambda^{(\varphi)}$, change at Q on $C_o^{(\varphi)}$ to the N distinct eigenvalues $\lambda_b^{(1)}, \lambda_b^{(2)}, \ldots, \lambda_b^{(N)}$ of A, with multiplicities $n_1, n_2, \ldots, n_N$ and that these remain constant until $C_o^{(\varphi)}$ intersects the shock line $\mathscr{D}$. Here, as before, we assume that $\lambda_b^{(i)}$ to be ordered so that $\lambda_b^{(1)} > \lambda_b^{(2)} > \ldots > \lambda_b^{(N)}$, and that

$$\sum_{i=1}^{p} m_i = \sum_{i=1}^{N} n_i = n.$$

Now to the left of $\mathscr{D}$ in $\mathscr{R}_u$ the eigenvalues of A and their associated eigenvectors are assumed to be continuous functions of x, t and U. Thus the change of multiplicity can only come about due to the coalescence of some of the distinct zeros of the characteristic determinant of A, and by the converse process. Because of the continuous dependence of the zeros on x,t,U and the fact that bifurcation occurs at Q on $C_o^{(\varphi)}$ after which the multiplicities remain constant there will, in general, be a curve $\mathscr{B}$ entering $\mathscr{R}_u$ from Q such that to its left the multiplicities are $m_1, m_2, \ldots, m_p$ and to its right they are $n_1, n_2, \ldots, n_N$ as in Fig.24 . From amongst the p characteristics of system (5.1) corresponding to $\lambda^{(1)}, \lambda^{(2)}, \ldots, \lambda^{(p)}$, there will be q_b that issue out from Q with increasing time and lie to the left of $\mathscr{B}$, with eigenvalues $\lambda^{(p)}, \lambda^{(p+1)}, \ldots, \lambda^{(p-q_b+1)}$ and corresponding multiplicities $m_p, m_{p-1}, \ldots, m_{p-q_b+1}$. Similarly, there will be q_b^* of the N distinct characteristics of system (5.1) that issue out from Q with increasing time and lie to the right of $\mathscr{B}$ with the respective eigenvalues $\lambda_b^{(1)}, \lambda_b^{(2)}, \ldots, \lambda^{(q_b^*)}$ and corresponding multiplicities $n_1, n_2, \ldots, n_{q_b^*}$.

The continuity of the eigenvalues implies that $\lambda^{(1)}(Q) = \lambda_b^{(1)}(Q)$, though the multiplicity of the characteristic curve comprising the wavefront trace will change from m_1 to n_1 on crossing Q. The wavefront trace will thus have a continuous tangent across Q on $C_o^{(\varphi)}$.

Let the characteristic curves issuing out from Q after bifurcation and lying to the right of $\mathscr{B}$ be denoted by $C_b^{(\psi)}$, $C_b^{(2)}, \ldots, C_b^{(q_b^*)}$, where $C_b^{(\psi)}$ is the wavefront trace corresponding to the eigenvalue $\lambda_b^{(\psi_b)} \equiv \lambda_b^{(1)}$ with

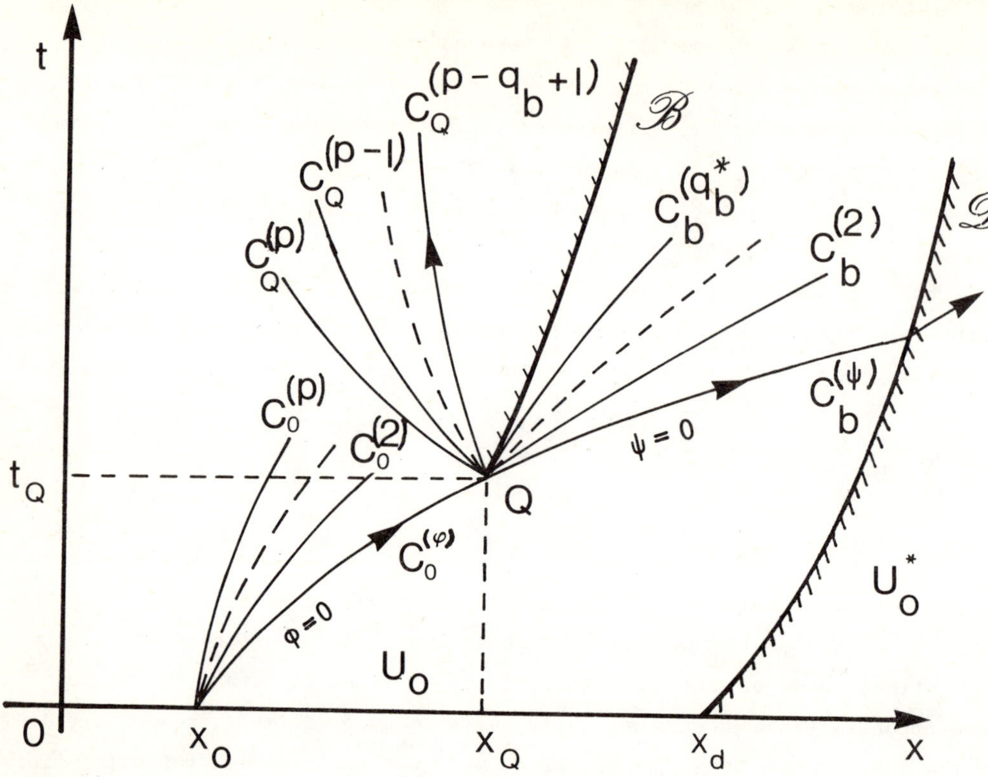

Fig. 24 Bifurcation on $C_o^{(\varphi)}$ at Q

multiplicity n_1. If $\psi(x,t)$ = constant along the family of characteristics in which $C_b^{(\psi)}$ is embedded, then $C_b^{(\psi)}$ must be a solution of the equation

$$C_b^{(\psi)} : \frac{dx}{dt} = \lambda_b^{(\psi)} \text{ where } \lambda_b^{(\psi)} = -\psi_t/\psi_x, \tag{5.81}$$

which passes through the point Q.

To parameterise this family in the same way as the $C^{(\varphi)}$ family of curves, along $t = t_q$ we shall set

$$\psi(x,t_q) = x - x_q, \tag{5.82}$$

for then $C_b^{(\psi)}$ corresponds to $\psi = 0$, with $\psi > 0$ ahead of $C_b^{(\psi)}$ and $\psi < 0$ behind it.

Working in terms of the new independent variables ψ and t'', where $t'' = t$, an analysis of the C^1 discontinuity propagation along $C_b^{(\psi)}$ will proceed in exactly the same manner to that along $C_o^{(\varphi)}$. Only the initial values of the

C^1 discontinuity quantities along $C_b^{(\psi)}$ will be different, and will need to be defined in terms of those along $C_o^{(\varphi)}$ to the left of Q at time $t = t_q$.

Hereafter the analysis of the problem closely parallels that of Section 5.5. The jump condition (5.4) is now no longer necessary, since the solution U is continuous across $\mathscr{B}$ at Q, so that only the manner of resolution of the incident C^1 discontinuity vector along $C_o^{(\varphi)}$ requires examination. Equations (5.45) and (5.47) are still valid in modified form, though with the absence of a shock, their right hand sides must now be equal. Hence we arrive at the result

$$\frac{1}{x_\varphi(t_q)}\left\{\Pi(Q) + \sum_{i=1}^{q_b} \Pi_i^{(R_b)}(Q)\right\} = \frac{1}{x_\psi(Q)}\sum_{i=1}^{q_{b^*b}} \Pi_i^{*(b)}(Q), \qquad (5.83)$$

where $\Pi_i^{(R_b)}(Q)$ is the initial value of the reflected bifurcation C^1 discontinuity vector across characteristic $C_Q^{(p-q_b+i)}$, and $\Pi_i^{*(b)}(Q)$ is the corresponding initial value of the transmitted bifurcation C^1 discontinuity vector along the characteristic $C_b^{(i)}$.

Analogous to equations (5.51) and (5.52), we have the results

$$\Pi_i^{*(b)}(Q) = \sum_{k=1}^{n_i} \delta_k^{(i)} r_{ob}^{(i,k)}(Q), \qquad (5.84)$$

and

$$\Pi_i^{(R_b)}(Q) = \sum_{k=1}^{m_i} \epsilon_k^{(i)} r_o^{(i,k)}(Q), \qquad (5.85)$$

where $r_{ob}^{(i,k)}$ is the right bifurcation eigenvector appropriate to the known solution U_o that lies ahead of both $C_o^{(\varphi)}$ and $C_b^{(\psi)}$, corresponding to the eigenvector $\lambda_b^{(i)}$ occurring with multiplicity n_i. Here, as in Section 5.5, the reflected bifurcation characteristics are referred to Q as origin and are parameterised in the obvious manner.

Defining the number N_b by the relationship

$$N_b = \sum_{i=1}^{q_{b^*}} n_i + \sum_{j=1}^{q_b} m_j, \qquad (5.86)$$

we see that (5.83) which implies n inhomogeneous algebraic equations for the N_b constants $\delta_1^{(1)}, \delta_2^{(1)}, \ldots, \delta_m^{(q^*b)}, \epsilon_1^{(1)}, \epsilon_2^{(1)}, \ldots, \epsilon_{mq_b}^{(q_b)}$ will have a solution only when $N_b = n$. This will be called the normal bifurcation case and we will assume it to apply. Once the constants are known the vector $\Pi_1^{*(b)}(Q$ can be found and the analysis of the behaviour of the solution along the bifurcation wavefront trace $C_b^{(\psi)}$ can proceed as before. In conclusion, we observe that if, for integer i, it is true that $\delta_1^{(i)} = \delta_2^{(i)} = \ldots = \delta_{n_i}^{(i)} = 0$, then at Q the system (5.1) is $\lambda_b^{(i)}$ bifurcation wave suppressive. Correspondingly, if, for some integer i, it is true that $\epsilon_1^{(i)} = \epsilon_2^{(i)} = \ldots = \epsilon_{m_i}^{(i)}$, then at Q the system (5.1) is $\lambda^{(p-q_b+i)}$ bifurcation wave nonreflective.

5.9 Wavefront Propagation in $\mathbb{R}^3 \times t$

Before concluding the development of the general theory of wavefront propagation in one space dimension and time, and prior to discussing two important applications, it is necessary to make brief mention of important contributions that have been made to the corresponding problem in $\mathbb{R}^3 \times t$. The general approach that has been developed for these problems is based on the differential geometry of the propagating discontinuity surface. For the essential mathematical background we refer to the books by Thomas [123] and Lichnerowicz [124] and for a short review of the use of these methods in continuum mechanics we refer to the paper by Thomas [125]. In a valuable collected set of papers Coleman, Gurtin, Herrerar and Truesdell [126] discuss the problems that arise when waves propagate in dissipative materials, for which the paper by Truesdell provides the mathematical background. The definitive source for continuum mechanics is the major review by Truesdell and Toupin [127]. An alternative source for material is to be found in the book by Eringen and Suhubi [54]. Applications of these same ideas to mathematical physics are to be found in the memoir by Boillat [128] and amongst his more recent papers as, for example, references [129,130]. See also Friedlander [131].

5.10 C^1 Wavefront Propagation in Shallow Water

The analytical approach that has been developed in this chapter is most readily illustrated by application to problems connected with the propagation of finite amplitude waves in shallow water. The governing equations for such problems have already been used to illustrate various ideas elsewhere in other chapters, and it is their simplicity of structure that makes them so convenient for our present purpose. In particular, the fact that only two equations are involved, thereby implying the existence of only a forward and a backward wave, obviates consideration of evolutionary conditions, whilst the fact that the associated eigenvectors are constant simplifies the details of the calculations.

The shallow water wave approximation is described by the equations in (1.16a,b) whose derivation is to be found in the reference work by Stoker [10]. For convenience we repeat the equations together with the definitions of the quantities involved:

$$u_t + uu_x + 2cc_x - H_x = 0, \tag{5.87a}$$

$$2c_t + 2uc_x + cu_x = 0, \tag{5.87b}$$

with u the x-component of fluid velocity and $c = \sqrt{gh}$ the local surface wave propagation speed, where h is the depth of the water, g is the acceleration due to gravity and $H(x) = gY(x)$, with $y + Y(x) = 0$ the equation of the seabed profile relative to the equilibrium surface of the water. To illustrate the effect of discontinuous coefficient matrices in the simplest possible way we consider a C^1 wave advancing into water at rest over a seabed profile which has a step change in depth from $y = -h_1$ to $y = -h_2$ across $x = x_p$. The fluid velocity will then remain continuous (and zero) across the advancing wavefront when it reaches $x = x_p$, though c and H will experience a discontinuous change across $x = x_p$. The geometrical configuration for this problem is shown in Fig. 25 and the details of the analysis are based on those given by Jeffrey in references [118,132,133].

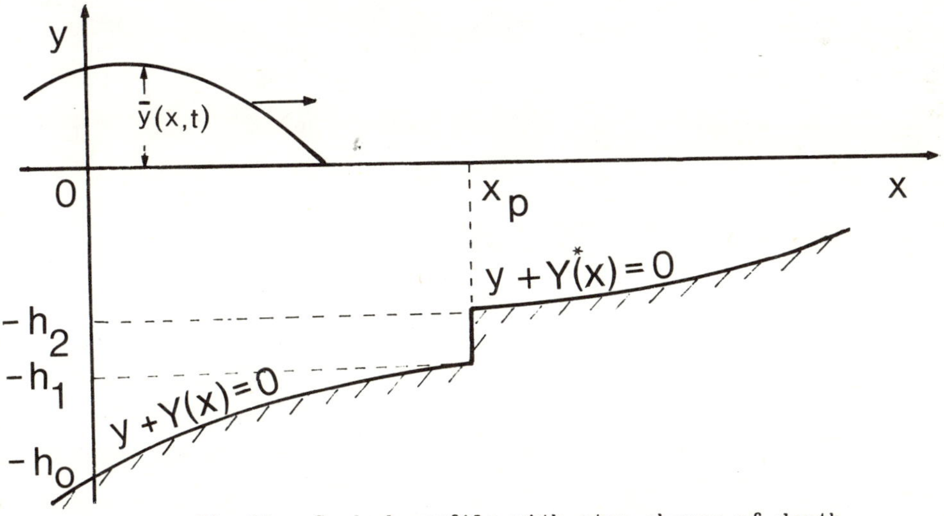

Fig.25 Seabed profile with step change of depth.

We denote by $\bar{y}(x,t)$ the finite amplitude disturbance to the water level relative to the equilibrium state at spatial coordinate x and time t, and assume that the wave which has a C^1 discontinuity at the origin is advancing to the right into the water at rest, with an initial acceleration

$$u_t(0-,0) = g\nu_o, \tag{5.88}$$

with ν_o a non-dimensional scale factor.

This form of initial condition may readily be reconciled with the periodic initial condition adopted by Stoker [79] who set $\bar{y}(o,t) = A\sin\omega t$. In terms of acceleration the periodic initial condition is equivalent to setting $\nu_o = A\omega/\bar{c}_o$, with $\bar{c}_o = \sqrt{gh_o}$ the surface wave propagation speed at the origin at which the water depth is h_o.

The discontinuity in the coefficient matrices will occur at $(x_p,0)$ on the initial line, showing that the line $\mathscr{D}$ must enter the upper half of the (x,t) plane from this point. Hence we have $\mathscr{I}_u = \{x|\ 0 < x < x_p;\ t = 0\}$ and, similarly, $\mathscr{I}_{u*} = \{x|x > x_p;\ t = 0\}$.

From (1.18a,b) the conservation form of (5.87) may be written

$$\frac{\partial F}{\partial t} + \operatorname{div} G = 0, \tag{5.89a}$$

with

$$F = \begin{bmatrix} u \\ \\ c^2 \end{bmatrix} \quad\text{and}\quad G = \begin{bmatrix} \tfrac{1}{2}u^2 + c^2 - gY \\ \\ uc^2 \end{bmatrix}. \tag{5.89b}$$

So, writing $\bar{c} = \sqrt{gh_1}$ and $\bar{c}^* = \sqrt{gh_2}$, the local surface wave speeds of propagation to the immediate left and right of the step change in depth, it follows from the jump condition (5.4) that across $\mathscr{D}$

$$\begin{bmatrix} \bar{c}^2 & -gh_1 \\ & 0 \end{bmatrix} - \begin{bmatrix} \bar{c}^{*2} & -gh_2 \\ & 0 \end{bmatrix} = \tilde{\lambda}\begin{bmatrix} 0 \\ \bar{c}^2 \end{bmatrix} - \tilde{\lambda}\begin{bmatrix} 0 \\ \bar{c}^{*2} \end{bmatrix}. \tag{5.90}$$

This yields the pair of equations

$$(\bar{c}^2 - gh_1) = (\bar{c}^{*2} - gh_2) \quad\text{and}\quad \tilde{\lambda}(\bar{c}^2 - \bar{c}^{*2}) = 0.$$

As h_1 and h_2 are arbitrary, the first equation offers no new information, for it simply defines $\bar{c}$ and $\bar{c}^*$ to the left and right of $\mathscr{D}$. However, the second equation shows $\tilde{\lambda} = 0$, so we conclude as would be expected that the shock line $\mathscr{D}$ is stationary in the (x,t)-plane and comprises the line $x = x_p$. Region $\mathscr{R}_u$ then lies to its left and $\mathscr{R}_{u*}$ to its right.

In $\mathscr{R}_u$ the matrix form of (5.87) is

$$U_t + AU_x + B = 0 \tag{5.91a}$$

with

$$U = \begin{bmatrix} u \\ c \end{bmatrix}, \quad A = \begin{bmatrix} u & 2c \\ c/2 & u \end{bmatrix}, \quad B = \begin{bmatrix} -gY_x \\ 0 \end{bmatrix}, \tag{5.91b}$$

while in $\mathscr{R}_{u*}$ the equivalent system is

$$U^*_t + A^* U^*_x + B^* = 0 \tag{5.92a}$$

with

$$U^* = \begin{bmatrix} u^* \\ c^* \end{bmatrix}, \quad A^* = \begin{bmatrix} u^* & 2c^* \\ c^*/2 & u^* \end{bmatrix}, \quad B^* = \begin{bmatrix} -gY^*_x \\ 0 \end{bmatrix} \tag{5.92b}$$

The eigenvalues of A are $\lambda^{(\pm)} = u \pm c$, when the corresponding left and right eigenvectors are found to be the respective constant vectors

$$\ell^{(\pm)} = [1, \ \pm 2] \quad \text{and} \quad r^{(\pm)} = \begin{bmatrix} 1 \\ \pm\frac{1}{2} \end{bmatrix}. \tag{5.93}$$

As only two waves are involved, for simplicity of notation the superscripts $\pm$ have been used in place of a numerical index. Corresponding expressions exist for equations in the starred region $\mathscr{R}_{u*}$. The families of characteristic curves are, correspondingly,

$$C^{(\pm)} : \frac{dx}{dt} = \lambda^{(\pm)} \ \text{in}\ \mathscr{R}_u, \tag{5.94}$$

and

$$C^{*(\pm)} : \frac{dx}{dt} = \lambda^{*(\pm)} \ \text{in}\ \mathscr{R}_{u*}, \tag{5.95}$$

where the initial conditions used to parameterise $C^{(+)}$ and $C^{*(+)}$ are those given in (5.20) and (5.23), with $x_0 = 0$ to allow for the C^1 discontinuity defining the wavefront that starts from the origin.

The numbers q and q* of Section 5.5 are each unity, as are the numbers m_i and m^*_i of (5.5). The C^1 discontinuity that will advance to the right into

$\mathscr{R}_{u*}$ from P, the point of intersection of the wavefront trace $C_o^{(+)}$ with $\mathscr{D}$ is associated with the eigenvalue $\lambda_o^{*(+)}$, whilst the reflected C_o^1 discontinuity returning from P into $\mathscr{R}_u$ will be associated with the eigenvalue $\lambda^{(-)}$.

For simplicity, let us consider the special case of a uniformly shelving seabed profile up to the point $x = x_p$ at which, after a step change in depth, the depth remains constant, so that

$$\begin{aligned} Y(x) &= h_o - mx \quad \text{for } x \le x_p, \\ Y^*(x) &= h_2 \quad \text{for } x > x_p, \end{aligned} \tag{5.96}$$

with $h_1 = h_o - mx_p$.

As it will be assumed that propagation to the right will be into still water we have

$$u_o = 0, \quad c_o(x) = \sqrt{g(h - mx)}\ , \tag{5.97}$$

when the wavefront trace $C_o^{(+)}$ itself will be given by integrating

$$C_o^{(+)} : \frac{dx}{dt} = u_o + c_o,$$

or by

$$\int_0^t d\tau = \int_0^x \frac{ds}{\sqrt{g(h - ms)}} .$$

So writing $\bar{c}_o = \sqrt{gh_o}$ gives

$$C_o^{(+)} : \frac{x}{h} = \frac{g}{\bar{c}_o} \left\{ t - \frac{mg}{4\bar{c}_o} t^2 \right\} . \tag{5.98}$$

In passing, we remark here that pre-multiplication of (5.91a) by $\ell^{(\pm)}$ gives

$$\ell^{(\pm)}(U_t + \lambda^{(\pm)}U_x) + \ell^{(\pm)}B = 0 \text{ along } C^{(\pm)},$$

which is equivalent to

$$u + 2c + mgt = \text{const along } C^{(+)}, \tag{5.99a}$$

and

$$u - 2c + mgt = \text{const along } C^{(-)}. \tag{5.99b}$$

The corresponding results for the starred equations in the constant depth region follow directly by replacing u, c by u^*, c^* and by setting m = 0. The expressions (5.99) are merely Riemann invariants, and they exist in the case of the inhomogeneous equations (5.87) only because of the special form of Y(x). We make no explicit use of these two results in what follows as they are implied by the transport equations.

Now the transport equations (5.32), (5.34) and (5.36) reduce in this case to [132,133]

$$\pi_{1t} + 2\pi_{2t} = 0, \tag{5.100a}$$

$$2(c_o)_x X + \pi_1 - 2\pi_2 = 0, \tag{5.100b}$$

$$X_t = \pi_1 + \pi_2, \tag{5.100c}$$

where the C^1 discontinuity vector Π has elements $\pi_1 = [u_\varphi]_{\varphi=0_+}^{\varphi=0_-}$ and $\pi_2 = [c_\varphi]_{\varphi=0_+}^{\varphi=0_-}$. From the initial condition (5.97) it follows that

$$(c_o)_x = -\left\{\frac{mg}{2}\right\} \{g(h - mx)\}^{-\frac{1}{2}}$$

or, from (5.98),

$$(c_o)_x = -\left\{\frac{mg}{2\bar{c}_o}\right\} \left(1 - \frac{mg}{2\bar{c}_o} t\right)^{-1}, \tag{5.101}$$

showing that at time t = 0,

$$c_{ox}(0) = -mg/2\bar{c}_o. \tag{5.102}$$

As $u \equiv 0$ across the wavefront trace $C_o^{(+)}$, the corresponding derivative of the function u_o ahead of the wavefront trace is

$$u_{ox}(0) = 0. \tag{5.103}$$

However, at time t = 0 the C^1 discontinuity forming the wavefront is assumed to be located to the immediate left of the origin, so that the quantity $u_x(0-,0) \neq 0$. Let us denote this initial value $u_x(0-,0)$ immediately behind the wavefront trace by $\tilde{u}_x$.

It is now possible to determine the initial values $\tilde{\pi}_1$ and $\tilde{\pi}_2$ of the elements of the vector Π that will be required in the determination of Π itself. By definition we have

$$\tilde{\pi}_1 = \pi_1(0) = u_\varphi(0)\Big|_{\varphi=0_-} - u_\varphi(0)\Big|_{\varphi=0_+},$$

or, denoting by $\tilde{x}_\varphi$ the value of x_φ at $t = 0$ immediately behind $C_o^{(+)}$, this may be written

$$\tilde{\pi}_1 = \tilde{u}_x/\tilde{x}_\varphi - (u_x/\tilde{x}_\varphi)_o$$

or

$$\tilde{\pi}_1 = \tilde{u}_x/\tilde{x}_\varphi - u_{ox}(0)/\tilde{x}_\varphi,$$

so that from (5.103) this simplifies to

$$\tilde{\pi}_1 = \tilde{u}_x/\tilde{x}_\varphi\,. \qquad (5.104a)$$

Similarly, we have

$$\tilde{\pi}_2 = \pi_2(0) = c_\varphi(0)\Big|_{\varphi=0_-} - c_\varphi(0)\Big|_{\varphi=0_+},$$

or in terms of $\tilde{c}_x$ the initial value of c_x immediately behind the wavefront trace

$$\tilde{\pi}_2 = \tilde{c}_x/\tilde{x}_\varphi - (c_x/\tilde{x}_\varphi)_o$$

leading to the result

$$\tilde{\pi}_2 = \tilde{c}_x/\tilde{x}_\varphi - c_{ox}(0)/\tilde{x}_\varphi.$$

Combining this last result with (5.102) then gives

$$\tilde{\pi}_2 = \tilde{c}_x/\tilde{x}_\varphi + mg/2\bar{c}_o \qquad (5.104b)$$

where in the last term we have used the fact that $\tilde{x}_\varphi = 1$. The unit divisor $\tilde{x}_\varphi$ is retained in the first term of (5.104b) to clarify the relationship between derivatives with respect to φ and x. The initial vector $\tilde{\Pi}$ is defined in terms of Π and (5.104a,b) as follows

$$\tilde{\Pi} = \Pi(0) = \begin{bmatrix} \tilde{\pi}_1 \\ \\ \tilde{\pi}_2 \end{bmatrix}. \qquad (5.104c)$$

This initial C^1 discontinuity vector will be required when (5.100a,b,c) are integrated as follows. Writing $s = c_{ox}$ and differentiating (5.100b) with respect to t gives

$$2s_t X + 2sX_t + \pi_{1t} - 2\pi_{2t} = 0$$

which, by virtue of (5.100a) and (5.100c), becomes

$$s_t X \;+\; s(\pi_1 + \pi_2) + \pi_{1t} = 0.$$

Substituting for X from (5.100b) then shows that

$$2s\pi_{1t} + (2s^2 - s_t)\pi_1 + 2(s_t + s^2)\pi_2 = 0, \tag{5.105}$$

but from (5.101) $s_t = -s^2$, so that (5.105) reduces to

$$\pi_{1t} + \frac{3}{2} s\pi_1 = 0.$$

Integration now yields

$$\pi_1 = \tilde{\pi}_1 \left(1 - \frac{mg}{2\bar{c}_0} t\right)^{-3/2}, \tag{5.106}$$

which when used with (5.100a) gives

$$\pi_2 = \tilde{\pi}_2 + \tfrac{1}{2}\tilde{\pi}_1 \left[1 - \left(1 - \frac{mg}{2\bar{c}_0} t\right)^{-3/2}\right]. \tag{5.107}$$

Now from the initial condition (5.20) we see X(0) = 0 so that setting $t = 0$ in (5.100b) shows $\tilde{\pi}_1 = 2\tilde{\pi}_2$. After some manipulation and use of results (5.104a,b) we arrive at the expressions

$$u_x(t) = -\left\{\frac{g\nu_0}{\bar{c}_0}\right\}\left\{\frac{\tilde{x}_\varphi}{x_\varphi(t)}\right\}\left\{1 - \frac{mg}{2\bar{c}_0} t\right\}^{-3/2}, \tag{5.108}$$

and

$$c_x(t) = -\left\{\frac{g}{2\bar{c}_0}\right\}\left\{\frac{\tilde{x}_\varphi}{x_\varphi(t)}\right\}\left\{\nu_0\left[2 - \left(1 - \frac{mg}{2\bar{c}_0} t\right)^{-3/2}\right] + m\right\}, \tag{5.109}$$

for the values of u_x and c_x immediately behind the wavefront trace $C_0^{(+)}$ at time t. When (5.106) and (5.107) are used with (5.100c) together with the definition of X in the form

$$X(t) \;=\; x_\varphi(t)\Big|_{\varphi=0_-} - x_\varphi(t)\Big|_{\varphi=0_+},$$

$$= x_\varphi(t) - x_{\varphi 0}(t),$$

after some manipulation we find

$$x_\varphi(t) = \tilde{x}_\varphi \left\{ \left[1 - \frac{mg}{2\bar{c}_o} t\right] - \left[\frac{g\nu_o}{\bar{c}_o}\right] t - \left[\frac{2\nu_o}{m}\right]\left[\left[1 - \frac{mg}{2\bar{c}_o} t\right]^{-\frac{1}{2}} - 1\right] \right\} . \tag{5.110}$$

In deriving this last result use has been made of the unstarred version of (5.62) and of $\lambda_o^{(+)} = u_o + c_o = c_o$, showing that

$$x_\varphi o(t) = \tilde{x}_\varphi h(t) = c_o(t)/c_o(0),$$

or

$$\tilde{x}_\varphi h(t) = c_o(t)/\bar{c}_o = \left[1 - \frac{mg}{2\bar{c}_o} t\right] . \tag{5.111}$$

Now to apply the techniques of this chapter to the transport of the C^1 discontinuity across $\mathscr{D}$ we now let $t = t_p$ be the time at which $\mathscr{D}$ is reached, when the quantities u_x, c_x and x_φ incident to the left of $\mathscr{D}$ at (x_p, t_p) follow by setting $t = t_p$ in (5.108) to (5.110). The time t_p itself follows from (5.98) by setting $x = x_p$.

It is at this stage in the argument that appeal must be made to (5.43) to relate the gradients across $\mathscr{D}$. However we first take note of the fact that since neither G nor G^* depend explicitly on t, that $\tilde{\lambda}(P) = (d\tilde{\lambda}/dt)_p \equiv 0$ and that $B^* \equiv 0$, the equation simplifies to

$$(\nabla_u G)_p (AU_x^{(R)} + B)_p = (\nabla_{u*} G^*)_p (A^* U_x^{*(T)})_p . \tag{5.112}$$

To the left of $\mathscr{D}$,

$$(\nabla_u F) = \begin{bmatrix} u & 2c \\ c^2 & 2uc \end{bmatrix},$$

and so as $u \equiv 0$ ahead of the wave

$$(\nabla_u F)_p = \begin{bmatrix} 0 & 2\bar{c} \\ \bar{c}^2 & 0 \end{bmatrix}, \tag{5.113}$$

with a corresponding starred expression existing to the right of $\mathscr{D}$ at P. Equations (5.45) and (5.52) combine to give

$$U_x^{(R)}(P) = U_x(P) + \frac{\beta}{x_\varphi(t_p)} r^{(-)}$$

or, explicitly,

$$U_x^{(R)}(P) = \begin{bmatrix} u_x(t_p) + \beta' \\ \\ c_x(t_p) - \frac{1}{2}\beta' \end{bmatrix}, \tag{5.114}$$

where $\beta' = \beta/x_\varphi(t_p)$. Using the fact that $U_{ox}^* \equiv 0$ and combining (5.47) and (5.51) then gives $U_x^{*(T)}(P) = \alpha r^{(+)}$, or,

$$U_x^{*(T)}(P) = \alpha \begin{bmatrix} 1 \\ \\ \frac{1}{2} \end{bmatrix}. \tag{5.115}$$

So, from (5.112), (5.114) and (5.115) we arrive at the two equations for α, β',

$$\bar{c}^2 u_x(t_p) + \beta' \bar{c}^2 = \alpha \bar{c}^{*2}$$

$$2\bar{c}^3 c_x(t_p) + mg\bar{c}^2 - \beta' \bar{c}^3 = \alpha \bar{c}^{*3},$$

which when solved give

$$\alpha = \left\{\frac{\bar{c}}{\bar{c}^*}\right\}^2 \left\{\frac{2\bar{c}c_x(t_p) + \bar{c}u_x(t_p) + mg}{\bar{c} + \bar{c}^*}\right\}, \quad \beta' = \left\{\frac{2\bar{c}c_x(t_p) - \bar{c}^* u_x(t_p) + mg}{\bar{c} + \bar{c}^*}\right\}. \tag{5.116}$$

It is interesting to notice that the magnitude $\beta = x_\varphi(t_p)\beta'$ of the reflected C^1 discontinuity vector will vanish, thereby causing $\mathscr{D}$ to be $\lambda^{(-)}$ wave non-reflective should it happen that

$$2\bar{c}c_x(t_p) - \bar{c}^* u_x(t_p) + mg = 0. \tag{5.117}$$

The case $x_\varphi(t_p) = 0$ is excluded since it is assumed that the breakdown in differentiability occurs beyond $\mathscr{D}$.

Consider the transmitted C^1 discontinuity vector Π^* with initial value $\Pi^*(P) = \alpha r_o^{(+)}$. As $h^* \equiv 1$, $\mu \equiv 1$ and $(\nabla_{u*}\lambda^{*(+)})_o = [1,1]$ it follows from (5.58) and (5.115) that the critical time t_c is given by

$$0 = 1 + \int_{t_p}^{t_c} \frac{3}{2}\alpha \, dt',$$

or by

$$t_c = t_p - \frac{2}{3}\left\{\frac{\bar{c}^*}{\bar{c}}\right\}^2 \left\{\frac{\bar{c} + \bar{c}^*}{2\bar{c}c_x(t_p) + \bar{c}u_x(t_p) + mg}\right\}. \tag{5.118}$$

Equivalently, in terms of x, this may be written as the critical distance

$$x_c = x_p - \frac{2\bar{c}^*}{\bar{c}}\left\{\frac{\bar{c}^*}{\bar{c}}\right\}^2 \left\{\frac{\bar{c} + \bar{c}^*}{2\bar{c}c_x(t_p) + \bar{c}u_x(t_p) + mg}\right\}, \tag{5.119}$$

and using results (5.108) to (5.110) with $t = t_p$, after some manipulation this is seen to be equivalent to

$$x_c = x_p\left\{1 + \frac{h_o}{3\nu_o(h_o - h_1)}\left\{\frac{h_2}{h_1}\right\}^{3/2}\left\{1 + \sqrt{\frac{h_2}{h_1}}\right\}\left[\left(\frac{2\nu_o x_p + h_o - h_1}{x_p}\right)\left\{\frac{h_1}{h_o}\right\}^{7/4} - \frac{2\nu_o h_1}{h_o}\right]\right\}. \tag{5.120}$$

In the special case $h_1 = h_2$, $h_o \neq h_1$ corresponding to a change of slope of the bottom at $x = x_p$ without a step change in depth this simplifies to give

$$x_c = x_p\left\{1 + \frac{2h_o}{3\nu_o(h_o - h_1)}\left[\left(\frac{2\nu_o x_p + h_o - h_1}{x_p}\right)\left\{\frac{h_1}{h_o}\right\}^{7/4} - \frac{2\nu_o h_1}{h_o}\right]\right\}. \tag{5.121}$$

As u_x becomes infinite at x_c it is reasonable to interpret x_c as the position at which the wave breaks. This expression is in agreement with the result due to Burger [134] which was obtained using an asymptotic method in the vicinity of the wavefront trace, though his result is not valid in the more general case discussed here when a step change in depth of water is involved.

When $h_o = h_1$, $h_1 \neq h_2$ we have the simplest case of a bottom which is flat on either side of a step change of depth. The result obtained from (5.120) by letting $h_o \to h_1$ is easily found to be

$$x_c = x_p + \frac{h_1}{3\nu_o}\left\{\frac{h_2}{h_1}\right\}^{3/2}\left\{1 + \sqrt{\frac{h_2}{h_1}}\right\}\left\{1 - \frac{3\nu_o x_p}{2h_1}\right\}, \tag{5.122}$$

and $x_c > x_p$ provided the last factor is positive. To deduce the meaning of this condition allow $h_2 \to h_1$, so that the bottom becomes completely flat. We obtain

$$x_c = \frac{2h_1}{3\nu_o} , \qquad (5.123)$$

which is a result derived elsewhere [79,132]. This shows that the last factor in (5.122) will be positive provided the wave breaks beyond x_p, which is an obvious condition.

The coefficient β', which determines the magnitude of the reflected discontinuity vector $\Pi^{(R)}$, reduces in the case of a flat bottom with a step to the simple factor

$$\beta' = \frac{g\nu_o}{\bar{c}_o}\left\{\frac{\sqrt{h_2} - \sqrt{h_1}}{\sqrt{h_2} + \sqrt{h_1}}\right\} , \qquad (5.124)$$

in which the expression in brackets is of particular interest. This factor was first found in a slightly different context by Lamb [Section 176,[135]] in connection with a simple analysis directed towards calculating the reflection coefficient of long waves at a step change of depth when a sinusoidal wave train propagates. Lamb defined the reflection coefficient to be the ratio of elevations of the reflected and incident waves, and found that it was equal to the factor $(\sqrt{h_2} - \sqrt{h_1})/(\sqrt{h_2} + \sqrt{h_1})$. When $h_1 = h_2$, so that no step occurs, then $\beta' = 0$.

The objection that the vertical component of fluid velocity is neglected in Lamb's analysis was raised by Bartholomeusz [136] who solved the full two dimensional problem to determine the correct reflection coefficient. Rather surprisingly, he found that when the wavelength of the incident wave was greater than the step height then Lamb's result appeared as the first approximation to this coefficient which was determined as the solution to an integral equation. This would suggest that even though the results obtained in the present Section are offered mainly to illustrate the application of the general theory that has been developed they may, nevertheless, provide a satisfactory first approximation to the results that would be obtained by the solution of the equivalent two dimensional shallow water wave problem.

The application of this approach to the passage of a wave over vertical walled objects on a flat seabed has been discussed by Jeffrey and Tin [137]. They took into effect the reflections at successive walls when determining the time and place of breaking in the sense just defined. The results employed a recurrence relation for the successive reflection and transmission coefficients and related them at the n-th step to the initial conditions at the start of the wave motion. This device reduces the calculation of the time and place of breaking of a wave over a stepped seabed to an equivalent, and trivial, problem for a wave over a flat seabed, the result of which is well known [10,132].

If a steady flow occurs over a specified seabed profile it is not difficult to solve the resulting transport equations numerically, and to determine from them the corresponding time and place of breaking of the wave. The required steady solution may be found analytically from equations (5.87), from which will follow u and c as functions of x together with the free surface profile which will no longer be horizontal. A series of calculations employing this approach will be published elsewhere by the author.

5.11 Smooth Fronted C^2 Waves in the Shallow Water Approximation

The final topic to be examined will be the extension of the previous wavefront analysis to the study of the propagation of smooth fronted C^2 discontinuities. A general and very detailed analysis, without the restriction to smooth fronted waves, has already been presented by Jeffrey and Inan [119], but the simpler study given here is based on an application made by Jeffrey [138] to smooth fronted waves in the shallow water approximation. Once again the simplicity of structure of the system of equations involved eases the task of analysis, as it did in Section 5.10. We refer to [138] for a discussion of the possible relevance of this problem to tsunami waves, and to the paper by Mader [139] for details of a numerical simulation of realistic tsunami waves that utilised the same shallow water approximation and compared the results with those obtained using marker and cell techniques.

We consider the propagation of a wave into shallow water at rest above an arbitrary smooth seabed profile. At the initial time, as in Section 5.10, the origin of the coordinates will be taken to be in the surface of the water at the front of the wave, with the x-axis lying in the surface of the water at rest, directed in the direction of propagation. The initial data will be chosen such that the free surface is everywhere continuous and once differentiable, but that a discontinuity in the derivative of the surface slope occurs across the origin. That is, we assume C^2 discontinuity exists across the wavefront trace. The geometrical configuration to be considered is shown in Fig. 26 . Let us assume an initial wave profile of the form $U(x,0) = \Phi(x)$, where because the wave is smooth fronted Φ is differentiable, but $\Phi''(x)$ has a bounded discontinuity across $x = 0$. The initial discontinuity

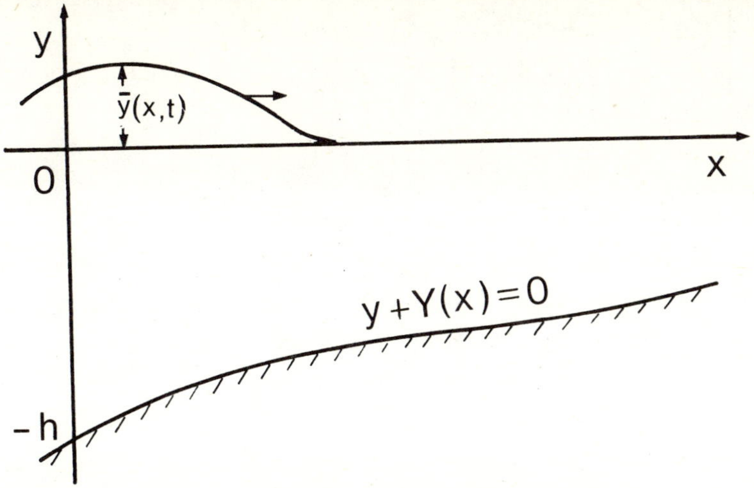

Fig. 26 Smooth fronted wave

propagating along the wavefront is determined once U(0,t) has been specified, though its magnitude is immaterial at this stage of the argument. Ahead of the wavefront trace the solution U_o corresponds to water at rest above the known seabed profile into which the wave is advancing. In this region $u = u_o = 0$ and $c = c_o(x) = H(x)^{\frac{1}{2}} = [gY(x)]^{\frac{1}{2}}$.

Although we have already seen that C^n discontinuities, and C^2 discontinuities in particular, propagate along characteristics, an alternative derivation of this fact is of interest.

It was first established by Douglis [38], and then in a more general manner by Lax [40], that discontinuities in first derivatives of the initial data propagate only along characteristics. This result was fundamental to the development of our study of propagating C^1 discontinuities and our object now is to use this work to prove directly that C^2 discontinuities, like C^1 discontinuities, propagate only along characteristics.

To see this we construct an extension of system (5.91) by differentiation. That is, we consider the enlarged system of equations comprising (5.91) together with the equations that result when (5.91) is differentiated with respect to x. Then, by the introduction of the column vector V with the four components u, c, u_x and c_x, it is a simple matter to show that this system is of the form

$$V_t + PV_x + Q = 0, \qquad (5.125)$$

with

$$V = \begin{bmatrix} u \\ c \\ u_x \\ c_x \end{bmatrix}, \quad P = \begin{bmatrix} A & \vdots & 0 \\ \cdots & \vdots & \cdots \\ 0 & \vdots & A \end{bmatrix}, \quad A = \begin{bmatrix} u & 2c \\ c/2 & u \end{bmatrix} \text{ and } Q = Q(V)$$

This first-order system has the same eigenvalues as (5.91) though their multiplicities are doubled. Because system (5.91) is hyperbolic and P has linearly independent sub-spaces, matrix P also has a full set of linearly independent eigenvectors. Thus (5.125) is also hyperbolic and has the same characteristics as system (5.91). The results of the papers by Douglis and Lax then assert that a C^1 discontinuity in column vector V of system (5.125) will propagate only along the characteristics of that system. That is, that a C^2 discontinuity in column vector U will propagate along $C_o^{(+)}$. This completes our proof.

(i) Transport Equations

In deriving the transport equations for the propagating C^2 discontinuity it is necessary to make use of the transport equations for the C^1 discontinuities that can exist across $C_o^{(+)}$. First, though, we remark that as the seabed profile is assumed to be smooth, it will follow that at all points ahead of the wavefront $c_o^2 = gY = H$ will be a differentiable function. The C^1 discontinuity transport equation analogous to (5.32) is

$$\ell^{(-)}(\Pi^{(1)} - X^{(1)}U_{ox}) = 0, \tag{5.126}$$

where $\Pi^{(1)} = [U_\varphi]_{\varphi=0_+}^{\varphi=0_-}$ denotes the C^1 discontinuity vector across the characteristic $C_o^{(+)}$ with equation $\varphi = 0$. The scalar

$$X^{(1)} = [x_\varphi]_{\varphi=0_+}^{\varphi=0_-} = x_\varphi\big|_{\varphi=0_-} - x_\varphi\big|_{\varphi=0_+}$$

similarly denotes the jump in the Jacobian x_φ across $C_o^{(+)}$.

Similarly, the transport equation analogous to (5.34) is

$$\ell^{(+)}(\Pi_{t'}^{(1)} + B_{ox}X^{(1)}) = 0. \tag{5.127}$$

The final equation connecting $X^{(1)}$ and $\Pi^{(1)}$ follows from (5.36) and is

$$\frac{dX^{(1)}}{dt'} = \pi_1^{(1)} + \pi_2^{(1)}, \tag{5.128}$$

where $\pi_1^{(1)}$ and $\pi_2^{(2)}$ are the components of $\Pi^{(1)}$ with

$$\pi_1^{(1)} = [u_\varphi]_{\varphi=0_+}^{\varphi=0_-} \text{ and } \pi_2^{(2)} = [c_\varphi]_{\varphi=0_+}^{\varphi=0_-} .$$

As before, equations (5.126) to (5.128) comprise a linear system for $\pi_1^{(1)}(t')$, $\pi_2^{(1)}(t')$, and $X^{(1)}(t')$ and have the trivial solution $\Pi^{(1)}(t') \equiv 0$ and $X^{(1)}(t') \equiv 0$. In the case of a smooth fronted wave we have the initial condition $\Pi^{(1)}(0) = 0$, but since the initial condition for φ given in (5.20) with $x_0 = 0$ shows that $X^{(1)}(0) = 0$, we see that the trivial solution is in fact the appropriate one for smooth fronted waves in respect of the propagation of C^1 discontinuities. Thus a water wave that starts smooth fronted, remains so, though we have still to determine how a C^2 discontinuity propagates. It is appropriate to remark here that $X^{(1)}(t') \equiv 0$ implies that the Jacobian x_φ is continuous across $C_0^{(+)}$, so that to find x_φ it is only necessary to use the fact that $u_0 = 0$ ahead of $C_0^{(+)}$ to deduce from (5.94) that the equation of a $C^{(+)}$ characteristic through the point $(x_0, 0)$ ahead of $C_0^{(+)}$ is

$$t = \int_{x_0}^{x} \frac{ds}{c_0(s)} . \tag{5.129}$$

Then, differentiaton of equation (5.129) with respect to φ coupled with the fact that t is independent of φ gives at a point (x, t) on this $C^{(+)}$ characteristic

$$x_\varphi = c_0(x)/c_0(x_0).$$

This result also follows, of course, from the unstarred version of equation (5.60). Now the wavefront trace in the (x, t)-plane is $C_0^{(+)}$, corresponding to $x_0 = 0$, so that the Jacobian on $C_0^{(+)}$ is given by setting $x_0 = 0$ in the above result when we find

$$x_\varphi = c_0(x)/\bar{c}_0, \tag{5.130}$$

where $\bar{c}_0 = c_0(0) = (gY(0))^{\frac{1}{2}} = \sqrt{(gh)}$ with $Y(0) = h$. Here we have chosen to express x_φ as a function of x rather than t', since by doing this it transpires that the resulting manipulations are simplified.

The form of argument that gave rise to equations (5.126) to (5.128) can again be used to derive transport equations for the vector $\Pi^{(2)}(t') = [U_{\varphi\varphi}]_{\varphi=0+}^{\varphi=0-}$ with components $\pi_1^{(2)}(t') = [u_{\varphi\varphi}]_{\varphi=0+}^{\varphi=0-}$ and $\pi_2^{(2)}(t') = [c_{\varphi\varphi}]_{\varphi=0+}^{\varphi=0-}$, which is the jump in the second derivative of U with respect to φ across $C_o^{(+)}$, and also for the scalar

$$X^{(2)}(t') = [x_{\varphi\varphi}]_{\varphi=0_+}^{\varphi=0_-}.$$

These are the C^2 discontinuities that are of concern in the shallow water approximation.

To introduce the required C^2 discontinuity quantities we first display the equation corresponding to (5.27), which is

$$x_\varphi u_{t'} - 2x_\varphi c_{t'} - 2cu_\varphi + 4cc_\varphi - gY_x x_\varphi = 0. \qquad (5.131)$$

After differentiation with respect to φ and differencing across $C_o^{(+)}$ this yields

$$-2c_{t'}X^{(2)} - 2c\pi_1^{(2)} + 4c\pi_2^{(2)} - gY_x X^{(2)} = 0, \qquad (5.132)$$

where use has been made of the fact that $u_o = 0 =$ constant ahead of $C_o^{(+)}$ and u_φ, c_φ are continuous across $C_o^{(+)}$. However, ahead of $C_o^{(+)}$ equation (5.131) becomes

$$-2x_\varphi c_{ot'} + 4c_o c_{\varphi o} - gY_x x_\varphi = 0, \qquad (5.133)$$

so that elimination of the term in Y_x between (5.132), (5.133) results in the first transport equation for the C^2 discontinuity

$$\pi_1^{(2)} - 2\pi_2^{(2)} + 2c_{ox}X^{(2)} = 0, \qquad (5.134)$$

which is analogous to (5.126).

Now the equation corresponding to (5.33) is

$$u_{\varphi t'} + 2c_{\varphi t'} - gY_{xx}x_\varphi = 0, \qquad (5.135)$$

and a further differentiation with respect to φ followed by differencing across $C_o^{(+)}$ yields the second transport equation for the C^2 discontinuity

$$\pi_{1t'}^{(2)} + 2\pi_{2t'}^{(2)} - gY_{xx}X^{(2)} = 0, \qquad (5.136)$$

which is analogous to (5.127).

The third and final transport equation for the C^2 discontinuity follows by taking the plus sign in (5.94), differentiating twice with respect to φ and differencing across $C_o^{(+)}$ to obtain

$$\frac{dX^{(2)}}{dt'} = \pi_1^{(2)} + \pi_2^{(2)} . \tag{5.137}$$

This result is analogous to equation (5.128).

(ii) Propagation Law for the C^2 Discontinuity

The variation of $\Pi^{(2)}$ along the characteristic $C_o^{(+)}$ forming the wavefront trace can be determined by solving equations (5.134), (5.136) and (5.137) in terms of some assumed initial condition for the C^2 discontinuity across $C_o^{(+)}$ at the time $t' = 0$. As already remarked, it is more convenient to work in terms of x as the independent variable along $C_o^{(+)}$ rather than t', and this can easily be accomplished by means of the operator on $C_o^{(+)}$ $d/dt' \equiv c_o(x)d/dx$, which follows as a direct consequence of the differential equation $dx/dt' = c_o(x)$ of $C_o^{(+)}$. If we denote the initial C^2 discontinuities present in $\pi_1^{(2)}(t')$ and $\pi_2^{(2)}(t')$ by $\tilde{\pi}_1^{(2)} = \tilde{\pi}_1^{(2)}(0)$ and $\tilde{\pi}_2^{(2)} = \pi_2^{(2)}(0)$, respectively, then it follows directly from equation (5.134) that

$$\pi_1^{(2)} - 2\pi_2^{(2)} + 2c_{ox}(0)X^{(2)}(0) = 0. \tag{5.138}$$

However, as $\varphi(x,0) = x$ we have $x_{\varphi\varphi} \equiv 0$ along the initial line, so that $X^{(2)}(0) = 0$. We thus conclude from (5.138) that

$$\tilde{\pi}_1^{(2)} = 2\tilde{\pi}_2^{(2)}. \tag{5.139}$$

Using equation (5.139) and the fact that $gY(x) = c_o^2(x)$ it is a simple matter to eliminate $X^{(2)}$ between equations (5.134), (5.136) and (5.137) and to arrive at the two results

$$\pi_1^{(2)}(x) = [u_{\varphi\varphi}]_{\varphi=0_+}^{\varphi=0_-} = \tilde{\pi}_1^{(2)}\{\bar{c}_o/c_o(x)\}^{3/2}, \tag{5.140}$$

and

$$\pi_2^{(2)}(x) = [c_{\varphi\varphi}]_{\varphi=0_+}^{\varphi=0_-}$$

$$= \frac{\tilde{\pi}_1^{(2)} c_o(x)c_{ox}(x)}{2}\left\{\left[\frac{1}{\bar{c}_o c_{ox}(0)}\right] + \frac{\bar{c}_o^{3/2}}{2}\int_0^x \left[\frac{c_{ox}^2(s) - c_{oxx}(s)}{c_o^{7/2}(s)c_{ox}^2(s)}\right]ds\right\}, \tag{5.141}$$

where, as before, $\bar{c}_o = \sqrt{(gh)}$. Expressions (5.140) and (5.141) are the required propagation laws along the wavefront trace $C_o^{(+)}$ for the C^2 discontinuity present in $\Pi^{(2)}$ at the initial time.

The behaviour of the horizontal component u of the water velocity at the surface is of greatest interest, so we now confine attention to equation (5.140). Ahead of $C_o^{(+)}$ the water is at rest, so that we have $u_o \equiv 0$, and thus $u_{\varphi\varphi}\big|_{\varphi=0_+} = 0$. It then follows from (5.140) that immediately behind $C_o^{(+)}$, $u_{\varphi\varphi}\big|_{\varphi=0_-}$ is simply

$$u_{\varphi\varphi}\big|_{\varphi=0_-} = \tilde{\pi}_1^{(2)}\{\bar{c}_o/c_o(x)\}^{3/2}. \tag{5.142}$$

To convert from derivatives with respect to φ to those with respect to x we first note that

$$u_{xx} = \frac{\partial}{\partial x}(u_\varphi/x_\varphi) = \varphi_{xx}u_\varphi + u_{\varphi\varphi}/x_\varphi^2$$

but since $u_\varphi = 0$ we arrive at the result

$$u_{xx} = u_{\varphi\varphi}/x_\varphi^2. \tag{5.143}$$

Combining result (5.143) with equations (5.130) and (5.142) and recognising from $\varphi(x,0) = x$ that $x_\varphi(0)\big|_{\varphi=0_-} = 1$, so that $\tilde{\pi}_1^{(2)} = \{u_{\varphi\varphi}(0)/x_\varphi^2(0)\}_{\varphi=0-} = \{u_{xx}(0)\}_{\varphi=0_-} = \tilde{u}_{xx}$, say, we finally obtain the result

$$u_{xx}\big|_{\varphi=0_-} = \tilde{u}_{xx}\left\{\frac{\bar{c}_o}{c_o(x)}\right\}^{7/2}. \tag{5.144}$$

We conclude from this that u_{xx} immediately behind the wavefront trace $C_o^{(+)}$ can only become infinite when the wave reaches the shoreline. This follows because $c_o(x) = (gY(x))^{\frac{1}{2}}$, showing that $c_o(x)$ only becomes zero, and hence $u_{xx}\big|_{\varphi=0_-}$ becomes infinite, when the depth $Y(x)$ of the water is zero. This may lead to the breaking of the wave on the shoreline, though whether or not this occurs depends on the subsequent motion of the shoreline itself. For an interesting account of the motion of the shoreline and of the possibility of certain waves climbing a beach without breaking, we refer to the paper by Carrier and Greenspan [141]. However, it is important to recall that until the shoreline is reached, u_x remains zero, since although $u_x = u_\varphi/x_\varphi$, we have already seen that $u_\varphi \equiv 0$. We refer to [138] for the details of how the initial values $\tilde{u}_{xx}$ and $\tilde{c}_{xx}$ may be related to the prescribed time variation of c at the origin. Further details of the propagation of general C^2 discontinuities can be found in [119] together with an application to plane shear waves in hyperelastic solids.

References

1. L. Hörmander. Linear partial differential operators. (Berlin: Springer, 1969).

2. S. Mizohata. The theory of partial differential equations. (London: Cambridge University Press, 1973).

3. B. Friedman. Principles and techniques of applied mathematics. (New York: Wiley, 1956).

4. S.G. Mikhlin. Linear equations of mathematical physics. (New York: Holt, Rinehart and Winston, 1967).

5. I. Stakgold. Boundary value problems of mathematical physics. Vols. I and II (New York: Macmillan, 1967).

6. R. Courant and D. Hilbert. Methods of mathematical physics. Vol. II Partial differential equations (New York: Wiley-Interscience 1962).

7. G.B. Whitham. Linear and nonlinear waves (New York: Wiley-Interscience, 1974).

8. A. Jeffrey and T. Taniuti. Nonlinear wave propagation. (New York: Academic Press, 1964).

9. R. Courant and K.O. Friedrichs. Supersonic flow and shock waves. (New York: Wiley-Interscience, 1948).

10. J.J. Stoker. Water Waves. (New York: Wiley-Interscience, 1957).

11. A. Jeffrey. Breakdown of the solution to a completely exceptional system of hyperbolic equations. <u>J. Math. Anal. and Applics.</u> 45 (1974), 375-381.

12. D.R. Bland. Plane isentropic large displacement simple waves in a compressible elastic solid. <u>J. Appl. Maths. and Phys</u>. (ZAMP) 16 (1965), 752-769.

13. E. Varley. Simple Waves in general elastic materials. <u>Arch. Rat. Mech. and Anal</u>. 20 (1965), 309-328.

14. W.D. Collins. One-dimensional nonlinear wave propagation in incompressible elastic materials. <u>Q.J. Mech. and Appl. Maths</u>. 19 (1966), 259-328.

15. A. Jeffrey and M. Teymur. Formation of shock waves in hyperelastic solids. Acta Mechanica 20 (1974), 133-149.

16. M. Klein and I.W. Kay. Electromagnetic theory and geometrical optics. (New York: Wiley-Interscience, 1965).

17. D.S. Mitrinovic. Analytic inequalities. (Berlin: Springer, 1970).

18. J. Hadamard. Lectures on Cauchy's problem in linear partial differential equations. (New York: Dover, 1952).

19. G. Hellwig. Partial differential equations. (New York: Blaisdell, 1964).

20. P. Garabedian. Partial differential equations. (New York: Wiley, 1964).

21. J. Schauder. Das Anfangswertproblem einer quasilinearen hyperbolischen Differentialgleichung zweiter Ordnung. Fund. Math. 24 (1934), 213-246.

22. J. Leray. Hyperbolic differential equations. Lecture Notes, Inst. Adv. Study, Princeton 1951-2.

23. P.D.Lax. On Cauchy's problem for hyperbolic equations and the differentiability of solutions of elliptic equations. Comm. Pure Appl. Math. 8 (1955), 615-633.

24. L. Garding. Le probleme de Goursat pour l'equation des ondes, 11th Skand. Math. Kongr. (1949), 255-258.

25. ———— Linear hyperbolic partial differential equations, with constant coefficients. Acta Math. 85 (1951), 1-62.

26. B.M. Ingersoll. An initial value problem for hyperbolic differential equations. Bull. Amer. Math. Soc. 74 (1948),1117-1124.

27. K.O. Friedrichs. Symmetric hyperbolic linear differential equations. Comm. Pure and Appl. Math. 7 (1954), 345-392.

28. K.O. Friedrichs. Symmetric positive linear differential equations. Comm. Pure and Appl. Math. 11 (1958), 333-418.

29. A. Jeffrey. The propagation of weak discontinuities in quasilinear symmetric hyperbolic systems. J. Appl. Math. and Phys. (ZAMP), 14 (1963), 301-314.

30. J. Bazer and O. Fleischman. Propagation of weak hydromagnetic discontinuities. Phys. Fluids 2 (1959), 366-373.

31. J. Bazer and W.B. Ericson. Hydromagnetic shocks. Astrophys. J., 129 (1959), 758-785.

32. P. Germain. Remarks on the theory of partial differential equations of mixed type and applications to the study of transonic flow.
Comm. Pure Appl. Math. 7 (1954), 117-144.

33. A.R. Manwell. The hodograph equations.
(Edinburgh: Oliver & Boyd, 1971).

34. F. John. Continuous dependence on data for solutions of partial differential equations with a prescribed bound.
Comm. Pure Appl. Math. 13 (1960), 551-585.

35. M.M. Lavrentiev. Some improperly posed problems of mathematical physics. (translation)
(Berlin: Springer, 1967).

36. K.O. Friedrichs. Nonlinear hyperbolic differential equations for functions of two independent variables.
Am. J. Math. 70 (1948), 555-588.

37. G. Courant and P.D. Lax. On nonlinear partial differential equations with two independent variables.
Comm. Pure Appl. Math. 2 (1949), 255-273.

38. A. Douglis. Existence theorems for hyperbolic systems.
Comm. Pure Appl. Math. 5 (1952), 119-154.

39. P. Hartman and W. Wintner. On hyperbolic differential equations.
Am. J. Math. 74 (1952), 834-864.

40. P.D. Lax. Nonlinear hyperbolic equations. Comm. Pure Appl. Math. 6 (1953), 231-258.

41. G.F.D. Duff. On wavefronts and boundary waves.
Comm. Pure Appl. Math. 17 (1964), 189-225.

42. F. John. Plane waves and spherical means applied to partial differential equations.
(New York: Wiley, 1955).

43. F. John. Non-admissible data for differential equations with constant coefficients.
Comm. Pure Appl. Math. 10 (1957), 391-398.

44. B. Riemann. Uber die Fortpflanzung ebener Luftwellen von enlicher Schwingungsweite.
Abh. Ges. Wiss. Gottingen 8 (1860), 43.

45. G.S.S. Ludford. On an extension of Riemann's method of integration, with applications to one-dimensional gas dynamics.
Proc. Camb. Phil. Soc. 48 (1952), 499-510.

46. A. Jeffrey. The evolution of discontinuities in solutions of homogeneous nonlinear hyperbolic equations having smooth initial data.
J. Math. and Mech. 17 (1967), 331-352.

47. G. Darboux. Lecons sur la Theorie Generale des Surfaces. Vol. 2. (Paris: Ganthier-Villars, 1887-1894). (1915)

48. A. Weinstein. On the wave equation and the equation of Euler-Poisson. Proc. Symp. Appl. Math., Wave Motion and Vibration Theory, Vol. 5. McGraw Hill (1954), 137-147.

49. L. Bers. Mathematical aspects of subsonic and transonic gas dynamics. Surveys in Applied Mathematics II. (New York: Wiley, 1958).

50. J.A. Smoller and J.L. Johnson. Global solutions for an extended class of hyperbolic systems of conservation laws.
Arch. Rat. Mech. Anal. 32 (1969), 169-189.

51. J.A. Smoller. A uniqueness for Riemann problems.
Arch. Rat. Mech. Anal. 33 (1969), 110-

52. R. Von Mises. Mathematical theory of compressible fluid flow. (New York: Academic Press, 1958).

53. B.L. Rozhdestvensky and N.N. Yanenko. Quasilinear systems and their application to the dynamics of gases (in Russian) (Moscow: NAUK, Physico-Mathematical Literature, 1968).

54. A.C. Eringen and E.S. Suhubi. Elastodynamics Vol. 1. Finite motion. (New York: Academic Press, 1974).

55. H.K. Rhee, R. Aris and N.R. Amundson. On the theory of multicomponent chromatography.
Phil. Trans. Roy. Soc. A267 (1970), 419-455.

56. P.D. Lax. Hyperbolic systems of conservation laws II.
Comm. Pure Appl. Math. 10(1957), 537-566.

57. E. Varley. Simple waves in general elastic materials.
Arch. Rat. Mech. Anal. 20 (1965), 309-328.

58. D. Parker. Bodies which adjoin regions of simple wave supersonic flow.
Quart. J. Mech. and Appl. Math. 18 (1965), 299-323.

59. E. Cumberbatch and E. Varley. Generalised self-similar flows.
J. Inst. Math. Appl. 2 (1966), 1-11.

60. Z. Peradzynski. Nonlinear plane k-waves and Riemann invariants. Bulletin de l'Academie Polonaise des Sciences, Serie des sciences techniques 19 No. 9 (1971), 59-66.

61. M. Burnat. The method of Riemann invariants for multi-dimensional non elliptic systems. Bulletin de l'Academie Polonaise des sciences, Serie des sciences techniques 17 No. 11-12 (1969), 97-104.

62. A. Jeffrey and M. Teymur. Formation of shock waves in hyperelastic solids. Acta Mechanica 20 (1974), 133-149.

63. L. Levine. Entropy and simple waves in multidimensional gas flow. Proc. Camb. Phil. Soc. 72 (1972), 299-302.

64. K.O. Friedrichs. Nichtlineare Differenzialgleichungen. Notes of lectures delivered at Gottingen. 1955.

65. P.D. Lax. Development of singularities of solutions of nonlinear hyperbolic partial differential equations. J. Math. Phys. 5 (1964), 611-613.

66. F. John. Formation of singularities in one-dimensional nonlinear wave propagation. Comm. Pure Appl. Math. 27 (1974), 377-405.

67. E.A. Coddington and N. Levinson. Theory of ordinary differential equations. (New York: McGraw Hill, 1955).

68. J. Nitsche. Uber Unstetigkeiten in der Ableitungen von Losungen quasilinearer hyperbolischer Differentialgleichungssysteme. J. Rat. Mech. Anal. 2 (1953), 291-297.

69. W.F. Ames. Discontinuity formation in solutions of homogeneous non-linear hyperbolic equations possessing smooth initial data. Int. J. Nonlinear Mech. 5 (1970), 605-

70. B.L. Rozhdestvensky, private communication, 1972.

71. A. Jeffrey. A note on the derivation of the discontinuity conditions across contact discontinuities, shocks and phase fronts. J. Appl. Math. Phys (ZAMP) 15 (1964), 68-71.

72. G. Boillat. A relativistic fluid in which shock fronts are also wave surfaces. Phys. Letters 50A (1974), 357-358.

73. A. Jeffrey. A note on the integral form of the fluid dynamic conservation equations relative to an arbitrary moving volume. J. Appl. Math. Phys.(ZAMP) 16 (1965), 835-837.

74. S. Chandrasekhar. Hydrodynamic and hydromagnetic stability. (London: Oxford, 1961).

75. A.G. Kulikovskiy and G.A. Lyubimov. Magnetohydrodynamics. (Massachusetts: Addison-Wesley, 1965); originally published by State Physics and Mathematics Press, Moscow 1962.

76. Shih-I Pai. Magnetogasdynamics and plasma dynamics. (Vienna: Springer, 1962).

77. I.G. Katayev. Electromagnetic shock waves. (London: Iliffe, 1966); originally published by Sovetskoye Rudio, Moscow 1963.

78. J. Bazer and W.B. Ericson. Nonlinear wave motion in magnetoelasticity. Arch. Rat. Mech. Anal. 55 (1974), 124-192.

79. J.J. Stoker. The formation of breakers and bores. Comm. Pure Appl. Math. 1 (1948), 1-87.

80. M. Froissart (Ed.). Hyperbolic equations and waves. (Berlin: Springer, 1970).

81. I.I. Glass. Shock waves and man. (Toronto: University of Toronto Institute for Aerospace Studies, 1974).

82. R.J. Knops (Ed). Symposium on non-well-posed problems and logarithmic convexity. Lecture Notes in Mathematics No. 316 (Berlin: Springer , 1973).

83. H.A. Levine. Some non-existence and instability theorems for solutions of formally parabolic equations of the form $Pu_t = -Au + (u)$. Arch. Rat. Mech. Anal. 51 (1973), 371-386.

84. R.J. Knops, H.A. Levine and L.E. Payne. Non-existence, instability and growth theorems for solutions of a class of abstract nonlinear equations with applications to nonlinear elastodynamics. Arch. Rat. Mech. Anal. 55 (1974), 52-72.

85. P. Germain. Shock waves, jump relations and structure. Advances in applied mechanics. [Edited by Chia-Shun Yih], Vol. 12 (New York: Academic Press, 1972 pages 131-144).

86. P.D. Lax. Weak solutions of nonlinear hyperbolic equations and their numerical computation. Comm. Pure. Appl. Math. 7 (1954), 159-193.

87. I.M. Gel'fand. Some problems in the theory of quasilinear equations. Am. Math. Soc. Trans. Series 2, 29 (1959), 295-381.

88. O. Oleinik. On discontinuous solutions of nonlinear differential equations. Am. Math. Soc. Trans. Series 2, 26 (1959),95-192.

89. B.L. Rozhdestvensky. Discontinuous solutions of hyperbolic systems of quasilinear equations.
Russian Math. Surveys. 15 (1960), 53-111; published by the London Mathematical Society.

90. S.N. Kruzhkov. Generalised solutions for the Cauchy problem in the large for nonlinear equations of first order.
Soviet Math. Dokl. 10 (1969), 785-788.

91. N.N. Kuznetsov. The weak solution of the Cauchy problem for a multi-dimensional quasilinear equation.
Math. Zametki 2 (1967), 401-410.

92. A. Douglis. Solutions in the large for multi-dimensional nonlinear partial differential equations of first order.
Am. Inst. Fourier de l'Univ. de Grenoble 15 (1965), 2-35.

93. I M. Gel'fand and G.E. Shilov. Generalised functions, Vol. 1 Properties and operations.
(New York: Academic Press, 1964).

94. E. Hopf. The partial differential equation $u_t + uu_x = \mu u_{xx}$,
Comm. Pure. Appl. Math. 3 (1950), 201-230.

95. J. Burgers. A mathematical model illustrating the theory of turbulence.
Advances in Applied Mechanics. Vol. 1.
(New York: Academic Press, 1948).

96. J.D. Cole. On a quasilinear parabolic equation occurring in aerodynamics.
Quart. Appl. Math. 9 (1951), 225-236.

97. P. Germain and R. Bader. Unicité des écoulements avec chocs dans la méchanique de Burgers. Office Nationale d'Etudes et de Recherches Aeronautiques Paris, 1953, pp 1-13.

98. E. Hopf. Generalised solutions of nonlinear equations of first order.
J. Math. Mech. 14 (1965), 951-972.

99. E. Hopf. On the right weak solution of the Cauchy problem for a quasilinear equation of first order.
J. Math. Mech. 19 (1969), 483-487.

100. P.D. Lax. Hyperbolic systems of conservation laws and the mathematical theory of shock waves.
SIAM Regional Conference Series in Applied Mathematics No. 11 (1973).

101. J. Glimm and P.D. Lax. Decay of solutions of systems of nonlinear hyperbolic conservations laws.
Am. Math. Soc. Memoir No. 101 (1970) pp 1-111.

102. O. Oleinik. Construction of a generalised solution of the Cauchy problem for a quasilinear equation of the first order by the introduction of vanishing viscosity.
Uspekhi Mat. Nauk. 14 (1949), 159-164, Am. Math. Soc. Trans. 33, 277-284.

103. C.C. Conley and J A. Smoller. Viscosity matrices for two-dimensional nonlinear hyperbolic equations.
Comm. Pure. Appl. Math. 23 (1970), 867-884.

104. P.D. Lax and B. Wendroff. Systems of conservation laws.
Comm. Pure Appl. Math. 13 (1960), 217-237.

105. P.D. Lax and B. Wendroff. Difference schemes for hyperbolic equations with a high order of accuracy.
Comm. Pure Appl. Math. 17 (1964), 381-398.

106. R.D. Ricthmyer and K.W. Morton. Difference methods for initial value problems. 2nd Ed.
(New York: Wiley-Interscience, 1967).

107. W.F. Ames. Numerical methods for partial differential equations.
(London: Nelson, 1969).

108. K.O. Friedrichs and P.D. Lax. Systems of conservation equations with a convex extension.
Proc. Nat. Acad. Sci. USA 68 (1971), 1686-1688.

109. P.D. Lax. Shock waves and entropy. Contained in: Contributions to nonlinear functional analysis. Edited by E.H. Zarantonello.
(New York: Academic Press, 1971), 603-634.

110. L.I. Sedov. Similarity and dimensional methods in mechanics.
(New York: Academic Press, 1959).

111. J. Glimm. Solutions in the large for nonlinear hyperbolic systems of equations.
Comm. Pure Appl. Math. 18 (1965), 697-715.

112. E. Conway and J. Smoller. Global solutions of the Cauchy problem for quasilinear first order equations in several space variables.
Comm. Pure Appl. Math. 19 (1966), 95-105.

113. R Burnside and A.G. Mackie. A problem in shock wave decay.
J. Australian Math. Soc. 5 (1965), 258-272.

114. R.M. Gunderson. The decay of a magnetohydrodynamic shock wave.
J. Appl. Math. Phys. (ZAMP) 19 (1968), 864-881.

115 E. Inan. Decay of weak shock waves in hyperelastic solids.
Acta Mechanica (in press)

116. A. Jeffrey. The development of jump discontinuities in nonlinear hyperbolic systems of equations in two independent variables.
Arch. Rat. Mech. Anal. 14 (1963), 27-37.

117. A. Jeffrey. The propagation of weak discontinuities in quasilinear hyperbolic systems with discontinuous coefficients. Part I - Fundamental theory.
Applicable Anal. 3 (1973), 79-100.

118. A. Jeffrey. The propagation of weak discontinuities in quasilinear hyperbolic systems with discontinuous coefficients Part II - Special cases and application.
Applicable Anal. 3 (1974), 359-375.

119. A. Jeffrey and E. Inan. The propagation of second order Lipschitz discontinuities in quasilinear hyperbolic systems with discontinuous coefficients.
Proc. Roy. Soc. Edinburgh, Sect. A., 74 (1975), 205-224.

120. E.S. Suhubi and A. Jeffrey. Propagation of weak discontinuities in a layered hyperelastic half space.
Proc. Roy. Soc. Edinburgh, Sect. A., (in press)

121. R.E. Meyer. On waves of finite amplitude in ducts.
Quart. J. Mech. Appl. Math. 5 (1952), 257-269.

122. H. Rhee and N.R. Amundson. An analysis of an adiabatic adsorption column: Part I, Theoretical development.
Chem. Eng. J. 1 (1970), 241-254.

123. T.Y. Thomas. Concepts from tensor analysis and differential geometry.
(New York: Academic Press, 1965).

124. A. Lichnerowicz. Elements de Calcul tensoriel.
(Paris: Lib. Armand Colin, 1958).

125. T.Y. Thomas. The general theory of compatibility conditions.
Int. J. Engng. Sci, 4 (1966), 207-233.

126. B.D. Coleman, M.E. Gurtin, I. Herrerar and C.W. Truesdell.
Wave propagation in dissipative materials.
(Berlin: Springer, 1965).

127. C.A. Truesdell and R.A. Toupin. The classical field theories. Vol. 3, part 1, of Encyclopedia of physics, edited by Flugge.
(Berlin: Springer, 1954).

128. G. Boillat. La Propagation des Ondes.
(Paris: Gauthier-Villars, 1965).

129. G. Boillat. Nonlinear electrodynamics: Lagrangians and equations of motion.
J. Math. Phys. 11 (1970), 941-951.

130. G. Boillat. A relativistic fluid in which shock fronts are also wave surfaces.
Phys. Letters 50A (1974), 357-358.

131. F.G. Friedlander. The wave equation of curved space-time.
(London: Cambridge, 1976).

132. A. Jeffrey. The breaking of waves on a sloping beach.
Z. angew. Math. u. Phys. (ZAMP) 15 (1964), 97-106.
See also addendum. Z. angew. Math. u. Phys. (ZAMP) 16 (1965), 712.

133. A. Jeffrey. On a class of non-breaking finite amplitude water waves.
Z. angew. Math. u. Phys. (ZAMP) 18 (1967), 57-65.
See also addendum. Z. angew. Math. u. Phys. (ZAMP) 18 (1967), 918.

134. W. Burger. A note on the breaking of waves on non-uniformly sloping beaches.
J. Math. Mech. 16 (1967) 1131-1142.

135. H. Lamb. Hydrodynamics, 6th Ed.
(London: Cambridge, 1932).

136. E.F. Bartholomeusz. The reflection of long waves at a step.
Proc. Camb. Phil. Soc. 54 (1958), 106-118.

137. A. Jeffrey and Saw Tin. Waves over obstacles on a shallow seabed.
Proc. Roy. Soc. Edinburgh, Sect. A., 71 (1973), 181-192.

138. A. Jeffrey. Smooth fronted waves in the shallow water approximation.
Proc. Roy. Soc. Edinburgh, Sect. A. 73A (1975), 107-116.

139. C.L. Mader. Numerical simulation of tsunamis.
J. Phys. Oceanogr. 4 (1974), 74-82.

140. E. Becker. Relaxation effects in gas dynamics.
Aeronautical Journal 74 (1970), 736-748.

141. G.F. Carrier and H.P. Greenspan. Water waves of finite amplitude on a sloping beach.
J. Fluid Mech. (1958), 97-109.

142. A. Jeffrey and V.P. Korobeinikov. Formation and decay of electromagnetic shock waves.
J. Appl. Math. Phys. 20 (1969), 440-447.

Index